Communication Services via Satellite

This second edition is dedicated to the memory of Derek Henry in recognition of his encouragement and long friendship.

Communication Services via Satellite

A Handbook for Design, Installation and Service Engineers

Second edition

Geoffrey E. Lewis

BA, MSc, MRTS, MIEIE

Butterworth-Heinemann Ltd
Linacre House, Jordan Hill, Oxford OX2 8DP

 PART OF REED INTERNATIONAL BOOKS

OXFORD   LONDON   BOSTON
MUNICH   NEW DELHI   SINGAPORE   SYDNEY
TOKYO   TORONTO   WELLINGTON

First edition published by BSP Professional Books Ltd 1988
Second edition published by Butterworth-Heinemann Ltd 1992

British Library Cataloguing in Publication Data
Lewis, Geoffrey E.
  Communication Services Via Satellite:
  Handbook for Design, Installation and
  Service Engineers. – 2 Rev. ed
  I. Title
  621.3825

ISBN 0 7506 0437 9

Printed and bound in Great Britain

Contents

Contents

Preface to the second edition

During the space of three short years since the first edition was completed, satellite systems alongside optical fibres, appear to have become the catalysts that were needed to speed up the integration process of global communications services. Now in the early 1990s, trans-Atlantic telephone calls are equally likely to be routed via satellite or submarine cable. Many of the established systems have now adopted Application Specific Integrated Circuit (ASIC) technology, with the advantages of improved reliability and reduced system size and cost, leading to a greater popularity. In spite of this development, discrete component circuits have been retained in a number of cases, because it is felt that with these it is easier to explain the principles involved.

In preparing this second edition, the opportunity has been taken not only to bring the previous work up to date, but also to extend it into the realms of some of the services that might be introduced before the end of the 1990s. Some of these proposals look like changing the concept of personal communications so completely, that saying 'Open Channel 9' to a pocket pen may no longer be science fiction.

Preface to the first edition

Until the introduction of satellites, communications by electrical and electronic means developed in a progressive and evolutionary manner. Now, in a relatively short time span, a new concept of signals delivery has become available – one that is not only capable of carrying all previous services, but also provides the means of distributing many new services, some that could not be supported by previous terrestrial means.

The success of a technological system usually depends upon its financial and social viability. The system costs ultimately being borne by the end user, he or she becomes the final arbiter on the success or failure of the project. It therefore falls upon the design engineer to provision an efficient and effective service. It is then the responsibility of the installation and service engineer to maintain this good quality of service.

The communications engineering fraternity has a history of providing for such developments, achieving this by continually updating skills and knowledge as the technology has developed. Satellite communications provide a whole new series of problems that will have to be solved by those responsible for the system support services.

To these personnel the book is offered, either as an aid for the self-training of practising engineers, or for communications engineering students following a more formalised training programme such as first degree, HNC/D or City and Guilds of London courses. Where it is anticipated that the reader will have some prior knowledge, reference is made to several standard works for added background information. The general philosophy has been to provide the reader with a description of how the systems signal are coded, modulated and demodulated, and processed, in order to create a good understanding of the way in which the system functions.

The book contains five chapters of related communications principles, provided to explain how and why certain system parameters are important. Also covered are as many actual systems as practicable, the approach in these chapters being largely based on system/block diagrams. There are two main reasons for this. Firstly, circuit diagrams for such rapidly developing technology are soon dated; and secondly, the expansion in the use of integrated circuits. These are closely related to sub-system blocks and their internal circuitry is often irrelevant, particularly from the service aspect. Engineers called upon to service such systems will be equipped with manufacturers' circuit diagrams. This book is intended to give engineers a thorough understanding of the system operation.

The systems design engineer needs to know how to assemble a cost effective and efficient system from the various standard units available, and the advanced student needs to understand similar concepts. For these people, a number of analyses and design rules have been included.

The design and construction of communications satellites and their methods of launch have been purposely omitted. These aspects are already well documented (1), (2).

REFERENCES

(1) Slater, J.N. and Trinogga, L.A. (1986) *Satellite Broadcasting Systems*. Chichester: Ellis Horwood Ltd.
(2) Rainger, P., Gregory, D., Harvey, R. and Jennings, A. (1985) *Satellite Broadcasting*. London: John Wiley & Sons.

Acknowledgements

During the research and writing period for this book, I corresponded with and met very many enthusiastic engineers. All provided me with much encouragement and many useful ideas, and provided valuable guidance. To all of those who read this work, I offer my grateful thanks in appreciation of their help. Such is the enthusiasm found among the practitioners of satellite communications that the task of writing this book has been made much easier and most stimulating.

Whilst I have acknowledged all sources of help in the References and Bibliography, I feel that I must express particular thanks to the following organisations and companies, and their personnel:

BICC Research and Engineering, London, U.K.: R. Grigsby, J.S. Buck.
British Telecom Research Laboratories, Martlesham Heath, U.K.: Dr N.D. Kenyon, M.D. Clark
Communications Systems Ltd, London, U.K.: J.L. Rose.
Electronic and Wireless World (Business Press International): G. Shorter, M. Eccles.
Eutelsat, Paris, France: P. Binet, G. De La Villetanet.
European Space Agency, Toulouse, France: G. Ferrand.
European Space Agency (Meteosat Operations), Paris, France: C. Honvault.
European Space Operations Centre, Darmstadt, West Germany: A. Robson.
European Space Research and Technology Centre, Noordwijk, Holland: N. Longdon.
Feedback Instruments Ltd, Crowborough, U.K.: P. Rowland-Smith, M.L. Christieson.
Independent Television Authority, U.K.: B.T. Rhodes, J.N. Slater.
INMARSAT, London, U.K.: D.M. Kennedy.
Marconi Communication Systems Ltd, Chelmsford, U.K.: P.A.T. Turrall.
Marconi International Marine Ltd, Chelmsford, U.K.: C. Riches.
Meteorological Office, Bracknell, U.K.: D.R. Maine, J. Turner.
Muirhead Office Systems Ltd, Beckenham, U.K.: K.A. Knowles.

Mullard Ltd, London, U.K.: M. Rope, B. Whale.

Multipoint Communications Ltd, Witham, U.K.: J.K. Player.

NEC Ltd, London, U.K.: S. Orme.

Plessey Semiconductors Ltd, Swindon, U.K.: J. Salter.

Polytechnic Electronics Ltd, Daventry, U.K.: M. Warman, E.L. York.

Rediffusion Engineering Ltd, Kingston-upon-Thames, U.K.: K.C. Quinton.

Satellite TV Antenna Systems Ltd, Staines, U.K.: S.J. Birkill, B.G. Taylor.

Scientific Atlanta Inc., Atlanta, U.S.A.: P.C. Bohana, K. Lucas.

Space Communications (Sat-Tel) Ltd, Northampton, U.K.: R. Crossley, J.P. Knowles.

Television (IPC Magazines Ltd), Contributors: H. Cocks, N. Harrold, H. Peters.

University of Kent at Canterbury, U.K.: Dr R. Collier, Dr G. McDonald, Dr E.A. Parker.

Video Systems (SVT) Ltd, Maldon, U.K.: G. Steele.

Wegener Communications Inc., Atlanta, U.S.A.: N.L. Mountain.

Chapter 1

The Basic Concepts

1.1 ADVANTAGES OF SATELLITE COMMUNICATIONS

Essentially, satellite radio communications systems are an extension of the relay systems that have been developed for terrestrial communications, with the difference that the receiver/transmitter, known in this case as the *transponder*, is now located in space. In spite of the high costs involved, the flexibility and advantages gained, allow satellites to make a valuable contribution to world-wide communications.

All the propagation parameters are well defined, well understood and capable of being accurately modelled mathematically. A single satellite can provide communications coverage for almost one third of the earth's surface, using much less radio frequency power than would be required for an equivalent terrestrial system. Even though the signal attenuation due to the long path lengths involved is high, it is fairly constant. A *signal fade margin* allowance of only about 3 to 5 dBs needs to be made to account for the variability due to local atmosphere and weather conditions. For the equivalent earth-bound system, an allowance of more than 30 dBs may be necessary.

To some extent, the high costs are due to the lack of service access to the satellite. This involves the use of on-board equipment redundancy and the provision of back-up satellites, to ensure continuity of service.

The high level of reliability of current technology is such that the operational lifetime of a satellite is now in the order of 15 years. When compared with terrestrial microwave relays and undersea cables, and taking into consideration the fact that in some areas of the world satellites are the only possible way of providing radio communications, the system becomes very cost effective.

1.2 ORBITS IN USE

Communications satellites occupy either an *equatorial*, an *elliptical*, or a *polar* orbit, as depicted in Fig. 1.1(a). *Geosynchronous* satellites are those

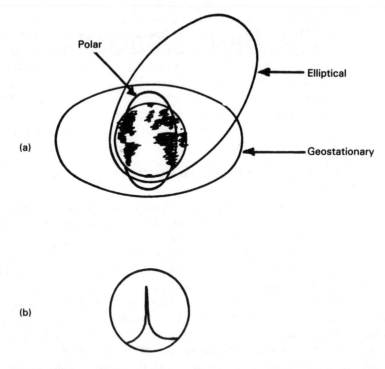

Fig. 1.1 (a) Orbits used for satellite communications. **(b)** Track of elliptical orbit over service area.

whose period of rotation is synchronised to that of the earth or some multiple of it. The *geostationary* orbit is a unique geosynchronous one, located over the equator. The satellite in geostationary orbit has a height and velocity such that it appears stationary to earth-bound observation. In this context, it is the earth's period of rotation relative to the fixed stars in space (the *sidereal* time) that is important. This is slightly less then the *solar* period and is approximately 23 hours, 56 minutes, 4.1 seconds (23.9345 hrs). The height above the earth's surface of 35,765 km and velocity of 3.073 km/sec (Appendix 1.1) required for the geostationary orbit was first calculated by the engineer and science fiction writer, Arthur C. Clarke (1) and presented in an article published in the October 1945 issue of *Wireless World*. This effectively set out the ground rules for satellite communications, so that the geostationary orbit is now named after Clarke in recognition of this work. The 'Clarke Orbit' is used to provide world-wide *point to point* (narrowcasting) and *point to multi-point* (broadcasting) services.

The elliptical-orbiting satellites operate with the earth's centre as one of the two focus points of an ellipse, which is inclined at a suitable angle to polar axis. Such an orbit has the advantage that the launch is achieved for a lower expenditure of energy. A typical satellite has an *apogée* (highest point) of about 35,600 km and a *perigée* (lowest point) of about 3,960 km above the earth's surface, and a period of slightly less than 12 hours. Around the apogée, the satellite appears *pseudo-stationary*, remaining within a beamwidth of less than ± 15° for more than 8 hours over the service area. Three satellites in such an orbit can therefore provide 24 hour-per-day service. These orbits, sometimes described as *super-synchronous*, are also known as *Molniya* orbits after the Russian communication satellites that used them. The track of such a satellite over its service area is shown in Fig. 1.1(b). With the crowding of the Clarke orbit that must occur as satellite communications expand, the elliptical orbits must become increasingly important during the 1990s.

The low polar orbits, with a typical height of about 850 km and a period of around 100 minutes, are valuable in navigation and weather forecasting. These satellites traverse the north and south poles, covering the earth's surface in a series of strips, to scan weather conditions around the world. Radio beacons are also carried for navigational purposes.

Personal mobile satellite communications can be established in the frequency band 1–2 GHz. Here the low level of propagation attenuation allows the use of small and simple antennas. By combining this technique with a number of *Low earth orbiters* (LEOs) equipped for inter-satellite communications and using modern integrated circuit technology, the concept of a world-wide *Personal Communication Network* (PCN) becomes a reality.

1.3 FREQUENCIES USED

Radio propagation is such that frequencies below some value are trapped within the ionosphere by reflection and refraction. This critical frequency varies during the 11-year sunspot cycle and with atmospheric conditions, but seldom exceeds 30 MHz. For this reason, only frequencies above about 100 MHz are used for space communications. The microwave frequencies used, which are shown in Fig. 1.2, are commonly known by their U.S. radar engineering designations.

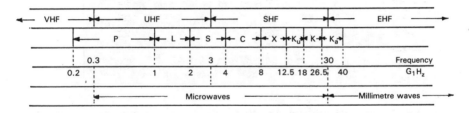

Fig. 1.2 Frequency bands and designations.

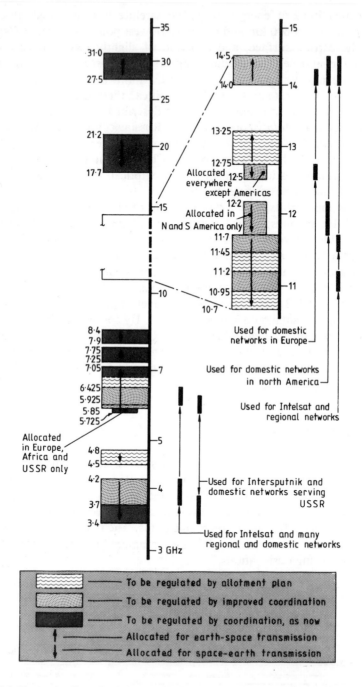

Fig. 1.3 Channels allocations due to WARC '85 (Courtesy of *Electronics & Wireless World* (2)).

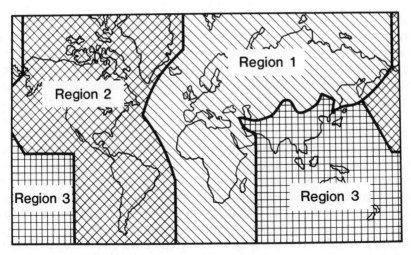

Fig. 1.4 Regions of the International Telecommunications Union (courtesy of IBA).

Table 1.1

Approximate Frequency GHz	Application
0.137	Weather satellite
0.145	Amateur
0.400	Satellite navigation
0.435	Amateur
0.714	Television–China
0.70 to 0.79	Television–U.S.S.R.
0.860	Television–India
1.269	Amateur
1.5 to 1.6	Marine–Inmarsat
1.7	Weather satellite
2.5 to 2.6	Television, etc.–Arabsat
40.5 to 42.5	Experimental (1986 onwards)
84 to 86	Experimental (1986 onwards)

Figure 1.3 shows how band sharing was internationally agreed at the World Administrative Radio Conference in 1985 (WARC '85) (2) for the various regions of the International Telecommunication Union (ITU), which are depicted in the map in Fig. 1.4. Other frequencies that may be used and the services provided on them are given in Table 1.1.

1.4   MAXIMISING THE AVAILABILITY OF THE FREQUENCY SPECTRUM

The methods used to regulate the exploitation of the frequency spectrum for terrestrial systems provided unfair advantage to those nations first in the field of radio communications. The result, in some cases, of almost haphazard development can be heard in nearly all wavebands. In certain parts of the spectrum, the level of mutual interference produced by transmitters is fast becoming intolerable. In a restricted sense, the frequency spectrum is a non-renewable resource. When a frequency is occupied by one transmitter, it is of little use to any other within the same coverage area.

World Administrative Radio Conferences (WARC '77, WARC '85) (2) were convened in order to manage the spectrum for space communications and minimise the effects on terrestrial systems. WARC '88 was planned so that the experience gained in the meantime could be consolidated. The problem of avoiding interference in space-based communications systems has a significant effect on the way in which geostationary satellites are managed. In these cases, mutual interference can be minimised by controlled allocation of orbit position, the use of frequency division, time division and code division multiple access techniques, and the use of various forms of signal polarisation.

The number of satellites that the Clarke Orbit will support depends upon the antenna pointing accuracy that can be achieved and maintained. Satellite operational lifetime is to some extent dependent upon the amount of fuel it can carry for station keeping. The accuracy required is within an arc of $\pm 0.1°$. A simple calculation shows that such a satellite can wander about its mean position within a sphere of about 62.4 km. Taking all these points into consideration, it is unlikely that spacings closer than about 2° can be tolerated using the technology that will be available during the first half of the 1990s.

Frequency division multiple access (FDMA) is a very common method where each transmitting station is allocated different carrier frequencies, as indicated in Fig. 1.5(a). Many stations may use the same transponder amplifiers simultaneously within the limit of the total channel bandwidth. Capacity allocation is simple and the system requires no complex timing or synchronism. The baseband signal is easily recovered using relatively simple and inexpensive equipment. The system can easily be computer modelled to minimise the risk of adjacent and co-channel interference from adjacent satellites.

The major disadvantage lies in the non-linearity of the transponder's travelling wave tube (TWT) high-power amplifier. Because several carriers are present simultaneously, intermodulation (IM) can arise. This is normally countered by *backing-off* the TWT power output, thus reducing the signal level received by the ground station.

Time division multiple access (TDMA) allows each earth station to be assigned a time-slot for transmission using the entire transponder resources as illustrated in Fig. 1.5(b). Each ground station transmitter uses the same carrier frequency within a particular transponder and transmits in bursts. As each burst must carry a destination address, the flexibility of interconnection is improved. Since only one carrier is present at any one time, the transponder's TWT can be operated near to saturation, for greater power output. There is no intermodulation or adjacent channel interference from the same satellite. This leads to an improved signal-to-noise ratio at the ground receiver. The cost of these improvements is the extra complexity of the equipment. However, the concept fits in well with digital processing

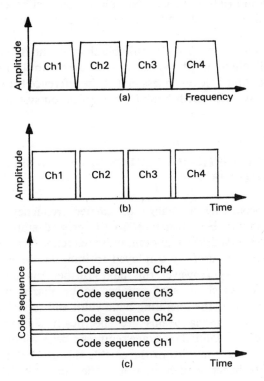

Fig. 1.5 (a) Frequency division multiplex. **(b)** Time division multiplex. **(c)** Code division multiplex.

developments, so this may not remain a disadvantage for long. A ground control station is necessary to maintain control and synchronisation of the system network. Through-put of data is better than FDMA by a factor of about two.

Code division multiple access is a method well suited to digital transmission. All users operate at the same carrier frequency and simultaneously use the whole channel bandwidth. A unique digital code is added to each digital transmission sequence and the intended receiving station equipped with the same *key*. (See Fig. 1.5(c).) By using *correlation detection*, the receiver extracts the wanted signal from the noise of all the other transmitters.

As the number of users increases, the system performance degrades gracefully, unlike normal digital systems whose performance *crashes* when overloaded. When the system is underused, there is an automatic improvement in the signal error margin.

Signal polarisation variation can permit *frequency re-use* to extend the occupancy of a waveband and, provided that the services are separated by sufficient distance, there will be minimum mutual interference. 'x' and 'y' linear and left and right-hand circular polarisation can be used in the same way as vertical, horizontal and circular polarisations have been used for terrestrial systems.

By using all these concepts, an equitable share of the frequency spectrum should be available to all who need to use it, even into the next century. In addition, there should be much less mutual interference between users than has been, the case with earth-bound communication systems.

1.5   SYSTEM CONSTRAINTS IMPOSED BY TRANSPONDER CHARACTERISTICS

Apart from producing a change of carrier frequency, the satellite's transponders should be *transparent* to the ground station receivers. In certain cases, *on-board* demodulation, noise reduction/error correction and remodulation can be usefully employed to improve the ground station's received signal quality. However, this makes further demands on the limited power available, which in any case has to be shared between all the transponders covering the operating bandwidth and services.

The down-link antenna beams are specifically shaped to cover the required service area. Global coverage requires a beamwidth of about 17.5° to cover a little over one third of the earth's surface. Relative aerial gains can be expressed in terms of beamwidth. The square of the beamwidth is approximately proportional to the antenna aperture. A typical *spot beam* might have a width of 4°. Thus the relative gains would be $(17.5/4)^2 = 19$ or 13 dB. Therefore, for the same signal level at the ground receiver, the global beam makes a much greater demand on the available power and limits the services available through the satellite. Higher effective radiated down-link power leads to smaller and lower-cost receiver antennae.

The transponder high-power stages commonly use *travelling wave tube* (TWT) amplifiers. These have the advantage of relatively low weight and high power efficiency. Figure 1.6 shows the shape of the amplitude and phase response of a typical TWT. When operated near to saturation, the non-linearity and *group delay* (rate of change of the phase response) will seriously distort amplitude modulated signals, and cause intermodulation when several carriers are present simultaneously. Such operation requires the TWT operating conditions to be *backed-off*, reducing the effective output power. Using frequency modulation, however, the TWT can be operated nearer to saturation for higher efficiency.

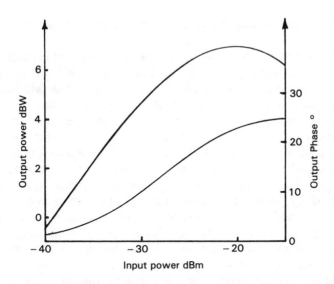

Fig. 1.6 Characteristics of a typical TWT amplifier.

1.6 CHARACTERISTICS OF MODULATION SYSTEMS

Before making a choice between the two basic systems, amplitude modulation and frequency modulation (AM and FM), two special cases need to be considered: those of analogue and digital baseband signals. The analogue signals will include voice telephony, audio and television (narrow and wideband). Digital signals may be derived from sources other than computer compatible systems. The case of digital signals derived from analogue ones also needs to be considered.

Taking analogue signals first, although from a frequency spectrum conservation viewpoint AM or single side band suppressed carrier (SSBSC) methods are more acceptable, they are not commonly used for satellite communications. A noise signal is strictly a random variation of amplitude, thus any receiver designed for AM will be responsive to noise. In addition, the amplitude varying nature of the AM signals requires that processing amplifiers should have linear characteristics, otherwise distortion will arise. This feature of the signal also imposes a varying demand on the power supply which must be capable of meeting the demand of peak modulation. In turn, this poses a serious problem for the satellite power supply and the TWTs.

By comparison, FM is a constant power method due to the constant amplitude nature of the signal. Thus there is no power supply problem and the high-power amplifiers can operate in Class C for maximum power efficiency. Also in the AM case, at least 50% of the available power is used to provide the carrier component which contains no information. In the FM case, all the transmitted power components carry information, so that for the same total power output FM produces a 3 dB more useful signal to a receiver.

It is demonstrated in Appendix 1.3 that the FM system of modulation has a signal-to-noise (S/N) ratio advantage over AM of $3M^2$, where M is the deviation ratio. In the case where M = 5, which is typical for wideband FM, the improvement is of the order of 19 dB. It is also shown here that noise is least troublesome at the lower baseband frequencies. This is due to differentiating action of the FM demodulator that reduces the noise power spectral density at the lower end. If a low-pass filter is added to follow the demodulator, this will cut the high-frequency noise component. The attenuation of the high-frequency baseband components can be compensated by boosting these in the transmitter in a complimentary manner before modulation. The use of this *pre-emphasis* and *de-emphasis*, shown in Fig. 1.7, that is not really available to AM systems can give rise to an average improvement of about 6 dB in the overall S/N ratio. (See Appendix 1.4.)

The FM receiver has another unique property, that of a *capture effect*. This is the ability of the receiver to lock on to the modulation of the stronger of two signals on the same or nearby frequency. Typically, a receiver can reject such signals that are only about 1 dB less than the wanted signal. Thus FM gives a superior rejection ratio for adjacent and co-channel interfering signals.

FM analogue signal processing can be further improved by the use of *companding*, where the dynamic range of the baseband signal is compressed before transmission and expanded in a complementary manner at the

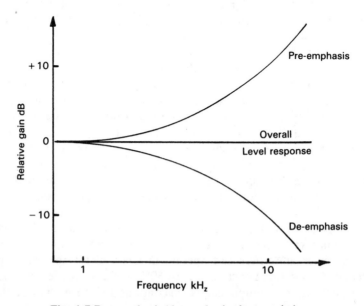

Fig. 1.7 Pre-emphasis/de-emphasis characteristics.

receiver. Not only does this improve the S/N ratio by perhaps as much as 9 dB, it also reduces the deviation ratio which reduces the modulated signal bandwidth and increases the satellite channel capacity.

Unlike AM detectors, FM demodulators can enhance the S/N ratio. The quality of the modulated signal is usually referred to by its *carrier-to-noise* (C/N) ratio, and after demodulation by its S/N ratio. Ignoring any small signal degradation, the AM detector's S/N ratio at the output will be equal to its C/N ratio. Within certain limitations depending on the deviation ratio, and FM demodulator can give a 15 dB improvement over AM (3), provided that the C/N ratio is above the 14 dB threshold. These features are displayed in Fig. 1.8. Recent circuit developments have produced FM demodulators with the threshold extended down to 3 dB, giving a further 11 dB improvement in the signal levels of FM. Figure 1.9 shows descriptively how the FM demodulator achieves the improvement derived in Appendix 1.3. The information signal bandwidth is halved and compressed back into its original baseband value, whereas the noise bandwidth is only halved.

The important feature of any signal processing must be to maximise the S/N ratio to obtain a high-quality analogue signal or a low error rate in a digital signal. This should be achieved whilst trying to transmit the maximum information within the available bandwidth, thus conserving the frequency spectrum. Carson's rule gives the approximate bandwidth B Hz for an FM signal as:

$$2(f_d + f_m) = 2(m + 1)f_m \qquad (1.1)$$

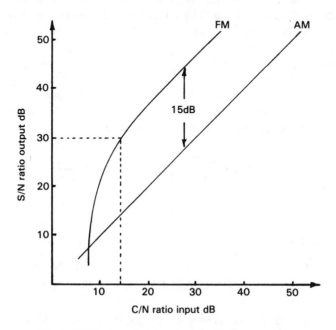

Fig. 1.8 FM demodulator threshold characteristics.

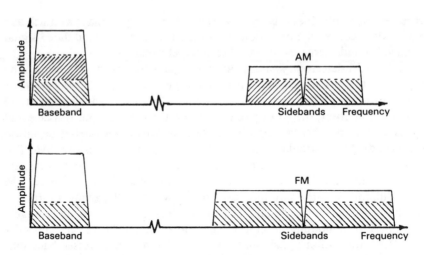

Fig. 1.9 Comparison of the effects of sideband noise between AM and FM.

where f_d is the frequency deviation in Hz, f_m is the modulating frequency in Hz, and $m = f_d/f_m$ the modulation index. This rule is illustrated in Fig. 1.10.

Shannon's rule for channel capacity using binary signalling is given by,

$$C = B \log_2 (1 + S/N) \text{ bits per second}$$

where C is the channel capacity, B is the bandwidth in Hz, and S/N is the S/N ratio. Shannon suggests that the capacity of a communications channel can be increased by increasing the bandwidth or the S/N ratio. Thus for an acceptable channel capacity, bandwidth can be traded for S/N ratio and vice versa. This is effectively what happens in the analogue FM case. Many coding and modulation combinations used for satellite communications operate within this trade-off.

It has thus been demonstrated that, under favourable conditions, FM can show a very significant S/N ratio improvement if the bandwidth is available, as is the case in satellite communications links.

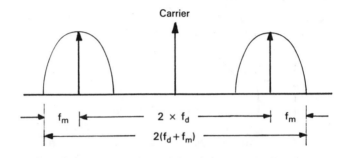

Fig. 1.10 Illustration of Carson's rule for FM bandwidth.

1.7 COMPANDING

Companding is the name given to the signal process that *compresses* the dynamic range of a signal before transmission and expands it in a complimentary manner in the receiver. This is achieved by using amplitude-dependent (non-linear) amplifiers with characteristics such as those of Fig. 1.11(a). The compressing amplifier in this example has a characteristic slope of 0.5. The receiver expanding amplifier therefore has a slope of 2, to give an overall unity response. Figure 1.11(b) shows how the process functions.

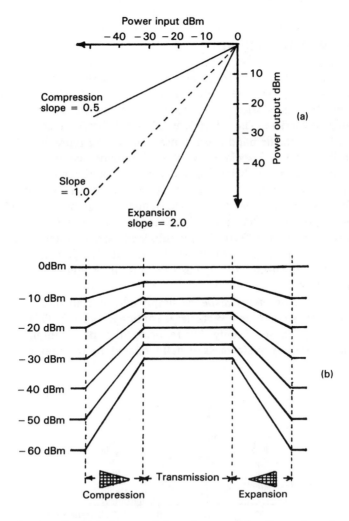

Fig. 1.11 Companding (a) characteristics of compression/expansion circuits; (b) overall system characteristic.

Signals above the level 0 dBm (1 μW) are assumed to be either non-existent or unaffected by companding. During the compression stage, the dynamic range is halved by increasing the relative amplitudes of the lower-level signals, which reduces the FM deviation necessary.

Assume that there is -30 dBm (1 μW) of noise present in the transmission media. This would have swamped the original low-level signal components, but now is only comparable with them. The expansion process at the receiver effectively depresses the low-level signal components (and the noise) to restore the dynamic range to its original value and improve the S/N ratio at the same time.

1.8 ENERGY DISPERSAL

The demands on the radio frequency spectrum are such that every possible technique must be used to minimise the interference between communication networks. Worst case conditions are usually caused by unmodulated carriers. The effects are reduced if the signal energy is spread evenly throughout the bandwidth in use. Figure 1.12 represents part of the frequency spectrum associated with the luminance component of a TV video signal, the energy being concentrated into small regions. Because the luminance signal rests at various dc levels for about 14% of the time (blanking levels), short bursts of single frequencies are produced. The burst repetition frequency is harmonically related to the line time base frequency. During these burst of effectively unmodulated carriers, interference can be caused to other services. The problem is particularly significant in C Band, where up and down-link TV signals can cause interference with terrestrial microwave links.

The technique of *energy dispersal* simply adds a triangular waveform to the video signal so that the FM carrier is continually varied. This spreads the spectral energy more evenly by reducing the peak amplitudes. The frequency of this waveform is usually related to the frame time base

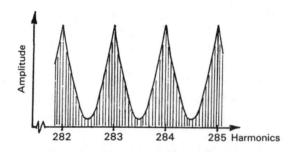

Fig. 1.12 TV luminance signal spectra.

frequency (25 or 30 Hz). It has been shown experimentally (4) that its amplitude should be such as to produce a peak-to-peak deviation of 600 KHz.

In at least one TV service (U.S.S.R., Gorizant), a dispersion frequency of 2 Hz is used, a frequency easily removed at the receiver by a good automatic frequency control (AFC) system. For higher frequencies, a video clamp circuit is necessary.

1.9 DIGITAL MODULATION SYSTEMS

As with analogue transmission, the main criterion for digital communications is still the accuracy of the received information. Therefore the aim must again to be to maximise the S/N ratio-bandwidth trade-off. It is shown (3) that any signalling pulse has an inverse *pulse width* to *pulse bandwidth* relationship. If the processing circuit bandwidth is less than the pulse bandwidth, distortion occurs and this leads to bit errors. To ease this problem, the data stream is usually prefiltered before modulation to produce a shape similar to $\frac{\sin x}{x}$ or related cosine pulses.

Whereas the C/N or S/N ratio was a valuable indicator of analogue signal quality, the ratio of energy per bit (E_b), per watt of noise power, per unit bandwidth (N_o) is often used for digital systems. (The ratio is dimensionless — joules = watts/sec.) The two ratios are related by;

$$\frac{E_b}{N_o} = \frac{C}{N} \times \frac{B}{R}$$

where $N_o = \frac{N}{B}$, $\frac{C}{N}$ is the C/N ratio in bandwidth B and R is the serial bit rate.

The *bit error rate* (BER) is the ratio of the number of bits received in error to the total number of bits transmitted per second. It is often typically expressed as $BER = e \times 10^{-3}$ per sec.

The most significant, and fairly predictable, problem is that of *white noise*, so called because its power spectral density (PSD) per unit bandwidth is constant. Statistically, such noise has a *Gaussian* or *normal* distribution, as shown in Appendix A1.5.

There are many ways of representing a binary message sequence with a series of pulses. The method chosen is always a compromise and two possible ways are shown in Fig. 1.13. The *non-return to zero* (NRZ) sequence maximises the energy per pulse and reduces the bandwidth needed. The *bi-polar* sequence has a spectrum with a zero dc component, a feature that is very useful if the signal is to be passed through ac circuits.

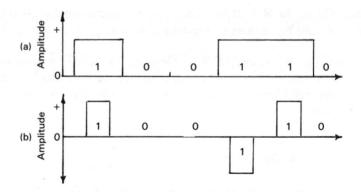

Fig. 1.13 (a) Non-return to zero pulse train. **(b)** Equivalent bi-polar sequence.

The three principal methods used to carry digital signals over radio frequency (RF) communication links are derived from the analogue concepts of amplitude, frequency and phase modulation.

On-Off Keying (OOK)

This is a limiting case of amplitude shift keying (ASK), where the carrier level is varied in discrete steps. In this case there are only two levels: on and off, 1 and 0. The RF pulse stream and frequency spectrum of such a signal is depicted in Figs. 1.14(a) and (b) respectively. It will be seen that the spectrum bandwidth is twice the baseband signalling frequency.

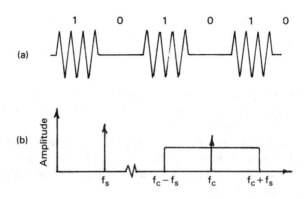

Fig. 1.14 (a) On-off keyed signal waveform.**(b)** Frequency spectrum.

It is explained in Appendix A1.5 that the probability of error (P_e) in a digital signal, in the presence of Gaussian noise, depends on the complementary error function (erfc) and a kind of S/N ratio. Assuming that the signal has a peak amplitude of A volts, that zeros and ones occur with equal probability and that the total noise power is N watts, then the error probability can be deduced from the S/N ratio:

$$\text{RMS signal amplitude} = \left(\frac{A/2}{\sqrt{2}}\right)$$

$$\text{RMS noise amplitude} = \sqrt{N}$$

$$\text{S/N ratio} = \left(\frac{A}{2\sqrt{2N}}\right)$$

$$P_e \text{ (OOK)} = \tfrac{1}{2} \text{ erfc.} \left(\frac{A}{2\sqrt{2N}}\right)$$

A similar expression based on the energy per bit ratio is:

$$P_e \text{ (OOK)} = \tfrac{1}{2} \text{ erfc.} \sqrt{\frac{E_b}{4N_o}}$$

Frequency Shift Keying (FSK)

This method involves switching the RF carrier frequency between two discrete values, each representing logic 0 or logic 1. Figure 1.15 shows the modulated wave at (a) and the frequency spectrum at (b). The upper and lower frequencies each have a spread of energy due to the baseband signalling frequency, so that it is apparent that FSK obeys Carson's rule (bandwidth $= 2(f_d + f_s)$). Again, the probability of error can be deduced using similar assumptions. The RF signal is present at all times so that its RMS voltage is $\dfrac{A}{\sqrt{2}}$

There are now two noise channels, one for each frequency so that the noise bandwidth is doubled. The RMS noise voltage is thus $\sqrt{2N}$.

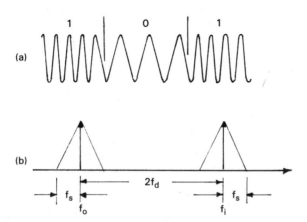

Fig. 1.15 (a) Frequency shift keyed signal waveform. **(b)** The frequency spectrum.

$$\text{S/N ratio} \quad = \frac{A/\sqrt{2}}{\sqrt{(2N)}}$$

$$P_e \text{ (FSK)} \quad = \tfrac{1}{2} \text{ erfc.} \quad \left(\frac{A}{2\sqrt{N}}\right)$$

Alternatively:

$$P_e \text{ (FSK)} \quad = \tfrac{1}{2} \text{ erfc.} \quad \sqrt{\frac{E_b}{2N_o}}$$

Phase Shift Keying (PSK)

This is essentially a single frequency method where the data stream causes the carrier phase to change. Figures 1.16(a) and (b) show the modulated waveshape and the phasor diagram. The signal is again present at all times, so there is only one signal channel and one noise channel. The S/N ratio is, therefore,

$$\frac{A/\sqrt{2}}{\sqrt{N}}$$

giving a probability of error

$$P_e = \tfrac{1}{2} \text{ erfc} \quad \left(\frac{A}{\sqrt{(2N)}}\right)$$

or, alternatively,

$$P_e = \tfrac{1}{2} \text{ erfc} \quad \sqrt{\frac{E_b}{N_o}}$$

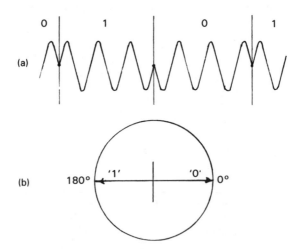

Fig. 1.16 (a) Phase shift keyed signal waveform. **(b)** The phasor diagram.

By comparison of these three terms for the probability of error, it can be seen that for the same value of P_e, PSK has a 3 dB advantage over FSK, which in turn has a 3 dB advantage over OOK. PSK also has the advantages of the narrowest bandwidth and the capability of expansion to carry even more data without an increase in bandwidth.

1.10 UP AND DOWN-LINK FREQUENCIES

Figure 1.3 indicates how the microwave part of the frequency spectrum has been allocated for particular services. In general, the up-link service is provided in the higher-frequency sub-band and the corresponding down link on a lower one. There are some economic and technical advantages in using this arrangement (4). It will be shown in chapter 3 that the gain of an antenna is proportional to the operating frequency and inversely proportional to its beamwidth. If the high band were used for the down link, any given antenna would have a narrower beamwidth, increasing the problems of maintaining antenna alignment as the satellite drifts around in space about its mean position. The added cost of a servo-controlled steering system would be required in compensation.

By using the high band for the up link, the extra gain can be usefully employed to make up for the extra path length attenuation, as these losses increase with frequency. With this arrangement, the overall system S/N ratio can be better managed. The large-dish ground transmitting station, will in any case require a servo-controlled tracking system, so that there is no added cost in this case(5).

REFERENCES

(1) Clarke, Arthur C. 'Extra-Terrestrial Relays'. *Wireless World*, Oct. 1945.
(2) Withers, David. 'Equitable Access to Satellite Communication' (review of WARC '85). *Electronics and Wireless World*, Dec. 1985.
(3) Schwartz, M. (1970) *Information Transmission, Modulation and Noise.* 2nd edn. (international student edition). London: McGraw-Hill Kogakusha.
(4) Griffiths, D.C. (1982) *IBA Technical Review No. 18.* Independent Broadcasting Authority.
(5) Ackroyd, Brian. (1990) *World Satellite Communications and Earth Station Design.* Oxford: BSP Professional Books.

Chapter 2
Noise and Interference

2.1 NOISE SOURCES

Noise in the communications sense is any spurious signal that tends to corrupt a wanted one. It is thus a destroyer of information and can be the ultimate limiting factor in a communications link. Noise may be *natural* or *artificial* (man-made) in origin. The former, which includes electromagnetic radiation from such as solar, galactic or thermal sources, is often described collectively as *sky noise*. It usually enters a communications system via a receiving antenna. Artificial noise is largely created by arcing contacts from sources such as vehicle ignition systems, commutator-type motors and generators, etc. It can include radiation from electrical supply systems, or even be carried by the mains wiring (mains-borne).

The effects of noise cannot be eliminated entirely, but by careful design and construction can be reduced to very acceptable levels.

Noise appears as random variation of signal voltages or currents, unrelated in phase or frequency. Such signals have a large peak-to-RMS ratio (*crest factor*) typically in the order of 4:1, so the annoyance factor is related to the noise power, or its mean square voltage or current.

A further type of noise that is usually described as interference is produced by other RF carriers present. In this case the noise tends to be periodic and regular in form. Nevertheless, it creates a major problem. Even a transmitter itself generates noise. Its non-linearities cause distortion which in turn produce unwanted radiation of harmonics. It also generates and radiates some thermal noise, so that even close to the transmitter there is a S/N ratio that is less than ideal.

2.2 SYSTEM NOISE FACTOR

Thermal noise, like white noise (described in chapter 1), has a statistically normal distribution. The annoyance factor of such noise is quantified (1) by the equation;

$$\overline{v^2}_n = 4kTBR \dots\dots\dots\dots\dots\dots\dots\dots\dots\dots\dots\dots\dots\dots \tag{2.1}$$

where

$\overline{v^2}_n$ is the mean square noise voltage
k   is Boltzman's constant 1.38×10^{-23} J/K
T   is absolute temperature K
B   is system bandwidth Hz and
R   is the source resistance

When a noise source is connected to a receiver, the worst case S/N ratio exists under maximum transfer of power conditions; that is, when the source and input resistances are equal. From Fig. 2.1, it can be seen that the maximum thermal noise power P_n, developed in the input load resistance, is given by:

$$(\overline{v_n}/2)^2/R = (\overline{v^2}_n/4R)$$
$$= 4kTBR/4R$$
$$= kTB \text{ watts or } kT \text{ watts/Hz} \qquad (2.2)$$

Thus the maximum available noise power is proportional to both temperature and bandwidth.

A receiver itself generates thermal noise power distributed throughout its stages. The measure of the input S/N ratio degradation so produced is described by the system *noise figure* or *factor* F, which is variously defined as;

(1) $\dfrac{\text{S/N ratio at system input}}{\text{S/N ratio at system output}}$

(2) $\dfrac{\text{S/N ratio of an ideal system}}{\text{S/N ratio of a practical system}}$

(3) $\dfrac{\text{Total noise power at output}}{\text{Noise power at output due to input alone}}$

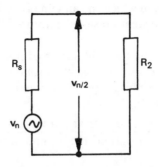

Fig. 2.1 Maximum available noise power conditions.

F may be quoted as a simple ratio or in decibels.

It is shown in Appendix 2.1 that the overall receiver noise factor F_0 is given by:

$$F_0 = F_1 + (F_2 - 1)/G_1 + (F_3 - 1)/G_1 G_2 \qquad (2.3)$$

where F_1, G_1, etc. are the individual noise factors and power gains of successive stages of the receiver.

Equation 2.3 shows the importance of low noise design for the system first stage, as this contributes most to the total degradation. The noise contributions of successive stages are smaller because these pass through fewer stages of amplification.

It should be pointed out here that an attenuator or other device that produces a signal level reduction has a noise factor numerically equal to its loss ratio L. Such a device should not therefore be used in the first stage, if it can possibly be avoided. The second term of equation 2.3 then becomes $(F_2 - 1)(1/L)$, so that the second stage makes a large contribution to the overall noise performance.

2.3 EQUIVALENT NOISE TEMPERATURE

It has been shown that the thermal noise power P_n, available from a resistor over a bandwidth of B Hz, is kTB watts, where T is the absolute temperature of the resistor. Equation 2.2 can be rearranged to give $T_e = P_n/kB$, where T_e is the effective temperature of the resistor that gives the same noise power. This concept can be extended to other noisy sources not necessarily associated with a physical temperature. For example, an antenna collects noise as random electromagnetic radiation and this may be associated with an *equivalent noise temperature*. If a noise power P_n watts is received by the antenna over a bandwidth of B Hz, then the antenna has an equivalent noise temperature, $T_a = P_n/kB$. Of course, T_a will depend very much upon the direction in which the antenna is pointed, thus showing that even the sky has an equivalent noise temperature. This is illustrated in Fig. 2.2.

Figure 2.3 shows how the third definition of noise factor is derived. The receiver is connected to a correctly matched source and assumed to be ideal (noise free). An auxiliary noise input T_r accounts for the equivalent noise temperature of the receiver.

$$F = \frac{\text{Total noise power at output}}{\text{Noise power at output due to input alone}}$$

$$= (GkT_s B + GkT_r B)/GkT_s B$$

$$= (T_s + T_r)/T_s = 1 + T_r/T_s$$

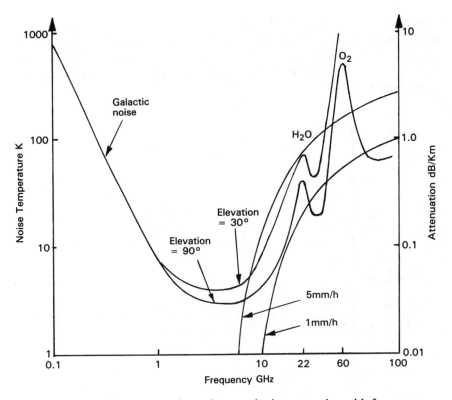

Fig. 2.2 Variation of sky noise and atmospheric attenuation with frequency.

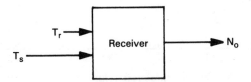

Fig. 2.3 Receiver noise factor.

or $F - 1 = T_r/T_s$.

By making this substitution in equation 2.3, an alternative expression for F_0 can be derived.

$$F_0 = 1 + T_1/T_sG_1 + T_2/T_sG_1G_2, \text{ etc.} \ldots E \, q.\tag{2.4}$$

2.4 COMPARISON OF NOISE FACTOR AND TEMPERATURE

When evaluating the performance of systems with low noise factors (less than about 5 dB), the scale for comparison becomes rather cramped. The graph of Fig. 2.4, which is a plot of equation 2.4 using a standardised source temperature of 290 K, illustrates how an improvement in F of 1 dB is equivalent to an improvement of noise temperature of about 130 K, thus showing the scale expansion that is available using the equivalent noise temperature concept.

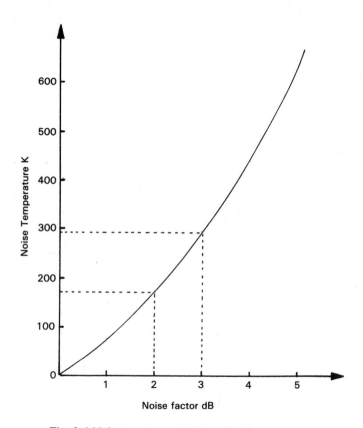

Fig. 2.4 Noise temperature/noise factor relationship.

2.5   ATMOSPHERIC EFFECTS

The atmosphere works in two ways to degrade the S/N ratio: by increasing the attenuation to lower the signal level, and by increasing the noise component. Figure 2.2 shows how both sky noise and rain effects are frequency dependent. Galactic noise falls significantly as frequency rises up to about 1 GHz. The noise level then remains fairly constant, depending upon the angle of elevation of the antenna, at about 4 K up to 12 GHz. This background noise is considered to be due to *black body* radiation of the expanding universe (the 'big bang' theory). The graph shows quite clearly why this range of frequencies is popular for satellite communications operations.

Both water (H_2O) and oxygen (O_2) molecules are capable of absorbing energy from electromagnetic radiation and then reradiating some of this as noise (scattering). This effect accounts for the two peaks in the curve at about 22 and 60 GHz.

Raindrops vary in size from a few microns to several millimetres in diameter, so storms can produce variation in absorption and scattering to degrade the S/N ratio. The effects of rain are also depicted in Fig. 2.2, light drizzle being rated at 1 mm/hr and moderate rain at 5 mm/hr. Except for the case of heavy rain, this form of attenuation is most noticeable above about 6 to 8 GHz and generally a 3 dB allowance needs to be made for this. However, if the transmitting and receiving stations are close to each other, as may be the case for television distribution, both the up and down links can be simultaneously affected. In this case the rain-fade margin will need to be doubled.

As the rain attenuation depends on the propagation path length through the storm, the effect increases as the angle of elevation of the antenna decreases. Assuming a storm depth of say 5 km and an elevation angle of 25°, the storm path length becomes $5/\sin 25° = 11.83$ km. Thus at the higher latitude receiving stations a greater margin must be allowed for.

If the rain is frozen as hail or snow, the attenuation falls to a relatively low level, perhaps 1% of that for liquid water (2). However, a special case exists. When the particles contain about 20% of liquid water, the attenuation can be about double that when the particles are completely melted.

The frequency ranges of approximately 18 to 20 GHz and 27.5 to 31 GHz (Ka band), can provide a further useful *window* for satellite communications (3). However, because of the rain effect, a fade margin in the order of 17 dB will need to be made.

2.6 INTERFERENCE WITH ANALOGUE SIGNALS

The major requirements of any receiver can be summarised as:

(1) To select the wanted frequency and reject all the unwanted signals that are present in any waveband.
(2) To recover the information from the modulated signal.
(3) To reproduce the information in a suitable manner.
(4) To carry out these functions without adding too much noise to the wanted signal.

The need to ensure that the S/N ratio remains as high as possible rests not only with the system designer but also, ultimately, with the service engineer.

It is therefore important to analyse the sources of interference to the systems.

Since the receivers in use for satellite reception are almost entirely double or triple conversion superhets, interference problems tend to be more complex than those found in the usual terrestrial equipment. Four basic forms of interference are due to *adjacent* and *co-channel*, *image* or *second channel* and *intermediate frequency* (IF) breakthrough.

Co-channel interference results from unusual propagation conditions and/or frequency reuse. Adjacent channel interference implies a lack of intermediate frequency (IF) selectivity within the receiver or discrimination at the antenna. Both are to some extent features of system design and band-planning, but when the problem arises in service, careful attention to the installation can often minimise the effects. Co-channel interference creates *whistles* on audio channels and *patterning* on television displays. In the latter case, high levels of interference can also give rise to *line tearing* because of the interfering synchronising (sync) pulses.

Adjacent channel interference is most likely to be apparent as sideband interference, as shown in Fig. 2.5(a), the higher baseband frequency components from each channel causing mutual interference.

Image or second channel interference is an unwanted feature of the superhet receiver concept. The wanted IF is the difference between the local oscillator and wanted carrier frequencies. Unfortunately, there are two carrier frequencies that give this result. These are shown in Fig. 2.5(b), where the *first* and *second* channels can be seen to be separated by twice the IF. Receivers of the double or triple conversion types thus make the problem of image channel rejection more complex.

Chosing a high-value IF at the design stage allows the RF stages ahead of the mixer to give good image channel rejection. However, this restricts the

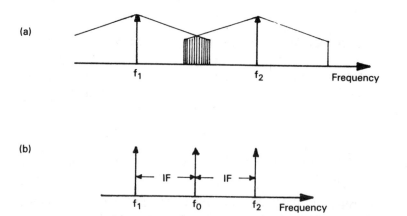

Fig. 2.5 (a) Sideband interference. **(b)** Second channel interference.

adjacent channel rejection properties of the IF stages, thus clearly indicating some of the design compromises that have to be made.

The superhet receiver can also suffer *IF breakthrough*. A strong signal present at the IF might break through into the IF stages where it will produce an effect similar to co-channel interference. For double or triple conversion receivers, this problem requires the use of efficient screening and filtering. From this can be seen the importance of choosing IFs from little-used parts of the frequency spectrum – another feature that calls for international agreement.

The ability of a receiver to reject these forms of interference is reflected in its *interference rejection ratios*. These are expressed in decibels as:

$$20 \log \left(\frac{\text{Wanted signal input voltage}}{\text{Interfering signal input voltage}} \right) \quad \text{(each producing the same output level)}$$

or:

$$10 \log \left(\frac{\text{Wanted signal output power}}{\text{Interfering signal output power}} \right) \quad \text{(each for the same input level)}$$

WARC '77 adopted a figure of merit for the direct broadcasting by satellite (DBS) television services. This was based on the *carrier to interference ratio* (C/I ratio) that produces *just perceptible* interference to the picture and is used to define the *service area* of a given channel. The *protection ratio* is quoted in decibels and, to a certain extent, is a subjective value (4). Protection levels of 31 dB for co-channel interference (CCI) and 15 dB for adjacent channel interference (ACI) were set as minimum requirements. In a practical situation, the *protection margin* is the value by which the actual C/I ratio exceeds these values. WARC '77 also proposed an *equivalent protection margin*, based on the summation of upper and lower adjacent channels and co-channel interferences, and the difference between this and the protection margin. Unless this value is positive, the level of interference will be greater than 'just imperceptible'.

It is common receiver practice to use a *low noise block convertor* (LNB) for the first stages. This wideband stage converts blocks of channels down to the first IF. Since there are several carriers present simultaneously, any non-linearity in the LNB will produce *intermodulation interference*. Some possible products of intermodulation are represented in Fig. 2.6.

2.7 SIGNAL TO NOISE RATIOS

The signal + noise to noise ratio is often used to describe the receiver sensitivity characteristic. This refers to the minimum input signal voltage required to produce a given output signal to noise ratio above the noise level for a specified degree of selectivity.

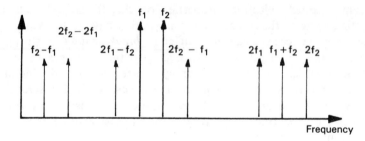

Fig. 2.6 Products of intermodulation.

The SINAD ratio takes into consideration the noise contribution produced by the distortion due to non-linear signal processing. It therefore relates to the signal + noise + distortion to noise + distortion ratio that results from processing a modulated carrier.

2.8 INTERFERENCE WITH DIGITAL CHANNELS

Noise and interference create similar problems for digital signals, distorting pulse shapes so that the pulses of adjacent symbols overlap, giving rise to *inter-symbol interference* (ISI) and creating bit errors at the detector stages. Digital processing by *slicing* and/or *sampling* can actually improve the S/N ratio, as shown by Fig. 2.7, but the price paid for this advantage is loss of accuracy of decoding under high noise conditions. The accuracy can be recovered by adding redundant bits to the message signal in the form of *error detection/correction codes*. However, this reduces the signalling rate by a factor:

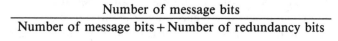

$$\frac{\text{Number of message bits}}{\text{Number of message bits} + \text{Number of redundancy bits}}$$

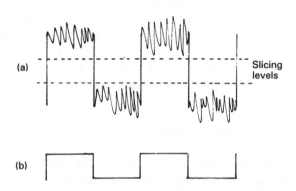

Fig. 2.7 (a) Noisy digital signal. **(b)** Regenerated signal.

The performance of analogue and digital signals under noisy conditions is compared in Fig. 2.8. As the input S/N ratio decreases, the output quality for the analogue system degrades gracefully, whilst the digital system is relatively unaffected until it suddenly *crashes*.

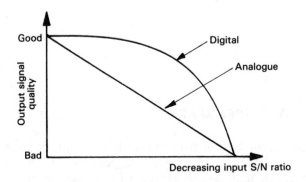

Fig. 2.8 Comparison of behaviour between digital and analogue signals in noise.

REFERENCES

(1) Connor, F.R. (1976) *Introductory Topics in Electronics and Tele-communication: Noise.* London: Edward Arnold Ltd.
(2) Wise, F. 'Fundamentals of Satellite Broadcasting'. *I.B.A. Technical Review* **11**, March 1979.
(3) Rose, J., Principal Consultant, Communications Systems Ltd, London: Private communication to author.
(4) Hopkins, D.K.W. *Interference Protection Ratios for C-MAC Vision.* I.B.A. Technical Report No. 126/83.

Chapter 3

Electromagnetic Waves, Propagation and Antennas

3.1 ENERGY IN FREE SPACE

Electromagnetic waves consist of energy in the form of *electric* (E) and *electromagnetic* (H) fields that are interdependent. The waves propagate through space due to an energy source, with the two fields acting at right angles to each other and mutually at right angles to the direction of propagation. Such waves occupy three-dimensional space and are said to be *orthogonal*. A very readable explanation of the way in which electromagnetic waves are forced to propagate, is given in reference (1).

·By convention, the *plane of polarisation* of these waves, is associated with the E field. Vertical and horizontal polarisations are used extensively in terrestrial communication frequency reuse systems. Here the E field is either vertical or horizontal in respect of the earth's surface.

3.2 POLARISATION AND THE TWO COMPONENTS OF THE E FIELD

Figure 3.1 shows the *Poincaré sphere*, a type of spherical quadrant, with centre O and the edges representing the three axes x, y and z. Consider O as the source of energy, generating an electric field, propagating in the arbitrary direction OP. At the surface of the sphere, the wavefront is perpendicular to the direction OP. Due to the angles θ and ϕ there will be components of the electric field in both horizontal (E_x) and vertical (E_y) directions, the total field E being given by $(E_x{}^2 + E_y{}^2)^{1/2}$. These two components, shown as being sinusoidal in Fig. 3.2, can be represented by:

$$E_x = E_1 \sin (\omega t) \text{ and}$$
$$E_y = E_2 \sin (\omega t + \delta)$$

where E_1 = amplitude of wave polarised in the x direction, E_2 = amplitude of wave polarised in the y direction, and δ = time-phase angle by which E_y leads E_x. In Fig. 3.2(b), δ is shown as $-90°$.

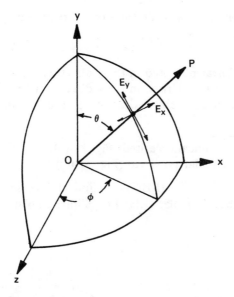

Fig. 3.1 The Poincare sphere and electric field vectors.

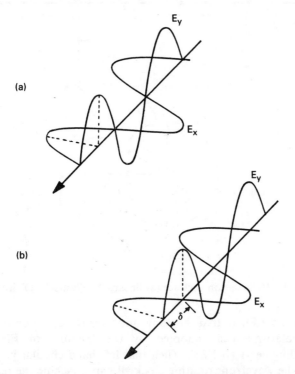

Fig. 3.2 (a) Linear polarisation. **(b)** Circular polarisation.

The polarisation of the wave depends on the relative magnitudes of E_1, E_2 and δ as tabulated below.

$\underline{\delta = 0}$

$E_1 = 0$, linear polarised in the y direction

$E_2 = 0$, linear polarised in the x direction (Fig. 3.3(a))

$E_1 = E_2$, linear polarised at 45° in the x,y direction

$\underline{\delta = \pm 90°}$

$E_1 \neq E_2$, elliptical polarised (Fig. 3.3(b))

$E_1 = E_2$, circular polarised (Fig. 3.3(c))

δ positive, left-hand circular (LHC) or elliptical polarised

δ negative, right-hand circular (RHC) or elliptical polarised

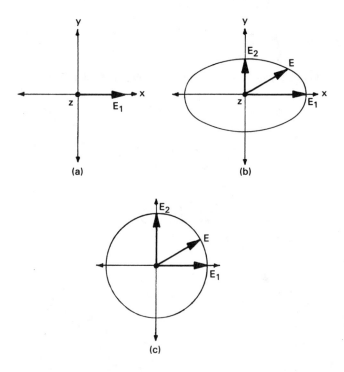

Fig. 3.3 Polarisation vectors: **(a)** linear; **(b)** elliptical; **(c)** circular.

It should be noted here that the convention for the direction of circular or elliptical polarisation as adopted by the Institute of Electrical and Electronic Engineers (IEEE) is such that left-hand circular polarisation is defined as the wavefront rotating clockwise approaching the receiver. This is opposite to that taken from classical optics.

3.3 PRIMARY PROPERTIES OF ANTENNAS

An antenna is essentially a transducer designed to obtain maximum transfer of energy from a transmitter into a communications medium or, alternatively, from the medium into a receiver. Because of the reciprocal properties of these devices, their characteristics can be measured or analysed in either the transmit or receive modes, as convenient, with the qualifications that the medium must be both linear and isotropic. Of the many types of antenna available, those based on the parabolic reflector are most commonly used for satellite communication purposes. The most important primary properties of antennas are: power gain, directivity, efficiency and equivalent noise temperature, for which there are various definitions.

Power gain is formally defined as:

$$G(\theta, \phi) = \frac{4\pi(\text{Power radiated per unit solid angle in direction } \theta, \phi)}{\text{Total power accepted by antenna from source}} \quad (3.1)$$

or alternatively by:

$$\frac{\text{Power radiated by practical antenna in preferred direction}}{\text{Power radiated in same direction from an isotropic antenna}} \quad (3.2)$$

with both antennas being supplied with the same power levels.

An isotropic antenna is a hypothetical device that radiates energy equally well in all directions. Such a device cannot be achieved in practice, but is a valuable theoretical reference resource with which to compare the properties of practical antennas.

The universal antenna constant:

$$G/A_e = 4\pi/\lambda^2 \quad (3.3)$$

is derived in Appendix A3.1, where:

G = power gain
A_e = effective antenna area
λ = Operational wavelength

Also $A_e = A\eta$, where:

A = physical area of antenna
η = coefficient of efficiency

It therefore follows that:

$$G = 4\pi A_e/\lambda^2 = \eta\pi^2 d^2 f^2/c^2 \quad (3.4)$$

where:

d = diameter of antenna
f = operational frequency
c = velocity of electromagnetic propagation

Thus showing that gain is dependent upon both dish diameter and operating frequency. Doubling either theoretically increases the gain by a factor of four, or 6 dB.

Directivity is the ability of an antenna to concentrate the radiated energy in a preferred direction in the transmit mode, or to reject signals that are received off-axis to the normal or *antenna boresight*. This gives rise to an amplifying effect on signals being received from the preferred direction.

The directivity is given by:

$$D(\theta, \phi) = \frac{4\pi(\text{Power radiated per unit solid angle in direction } \theta, \phi)}{\text{Total power radiated by antenna}} \qquad (3.5)$$

This property should be rotational symmetric, but this is not often achieved in practice.

The directivity for one practical antenna is shown in Fig. 3.4. The peaks and nulls are due to the phasor addition of signals components incident on different parts of the antenna surface. In some directions these components add, whilst in others phase cancellations occur, depending on the angles Θ and ϕ of Fig. 3.1. The side lobes which are produced chiefly by:

- reflector surface irregularities that give rise to scattering
- absorption losses at the reflector surface
- diffraction effects at the reflector edges
- the feedhorn *sees* a greater space than that occupied by the focal point thus producing an off-axis response

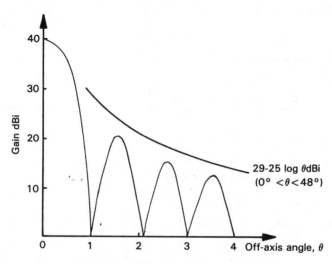

Fig. 3.4 Antenna directivity pattern.

These result in the antenna generating spurious emission or reception in unwanted directions, thus giving rise to interfering effects. To minimise this, the Consultative Committee for International Radio (CCIR) and the Federal Communication Commission (FCC) have adopted standards for geostationary satellite communications that ensures that the amplitude of side lobes will generally be below the level 29–25 log θ dBi (θ = the angle off-bore sight), as shown in Fig. 3.4.

These authorities have adopted slightly different standards to cater for satellite spacing at 2° and for varying European and American operational conditions.

CCIR	FCC
29–25 log θ dBi for $\theta<20°$	29–25 log θ dBi for $\theta<7°$
32–25 log θ dBi for $20°<\theta<48°$	8 dBi for $7°<\theta<9.2°$
<10 dBi for $\theta>48°$	32–25 log θ dBi for $9.2°<\theta<48°$
	10 dBi for $\theta>48°$

It is shown in reference (2) that the graphical form of Fig. 3.4 can be represented by the equation:

$$E_o = 2\lambda J_1((\pi d/\lambda) \sin \phi)/(\pi d \sin \phi)$$

where for $\theta = 90°$:

> J_1 = Bessel function of first order
> d = antenna diameter
> λ = operational wavelength
> ϕ = angle with respect to bore sight

Bessel functions, which were derived by F.W. Bessel (1784–1846) to solve Kepler's problem relating to the gravitational effects between three bodies in space, are similar in form to sinusoids of diminishing amplitude. They are usually presented in table or graphical form. Table 3.1 is a short extract from one such table.

Table 3.1

x	$J_1(x)$	
1	0.4401	
2	0.5767	
3	0.3391	} zero crossing
4	−0.0660	
5	−0.3276	
6	−0.2767	
7	−0.0047	} zero crossing
8	0.2346	
9	0.2453	
10	0.0435	} zero crossing
11	−0.1768	

By using linear interpolation to obtain approximation, the first three zero crossings occur at 3.84, 7.02 and 10.2. Thus the values that make $E_o = 0$ occur when:

$$J_1((\pi d/\lambda) \sin \phi) = J_1(3.84), J_1(7.02), J_1(10.2), \text{ etc.}$$

when:

$$(\pi d/\lambda) \sin \phi = 3.84, 7.02, 10.2, \text{ etc.}$$

so that $\sin \phi = 1.22(\lambda/d), 2.25(\lambda/d), 3.25(\lambda/d)$, etc.
Now since $\phi < 0.1$ rd. $\sin \phi \approx \phi$, so that:

$$\phi(\text{rd}) = 1.22(\lambda/d), 2.25(\lambda/d), 3.25(\lambda/d)... \text{ and}$$
$$\phi(^0) = 70(\lambda/d), 129 (\lambda/d), 186(\lambda/d) \qquad (3.7)$$

It is also shown in Appendix 3.2 that the half power (-3 dB) beamwidth is given by $57.3(\lambda/d)$.

The literature quotes two versions for the *efficiency* parameter. The one applying particularly to the transmit mode is the *radiation efficiency* (η):

$$= \frac{\text{Total power radiated by antenna}}{\text{Total power accepted by antenna from source}}$$

$$= G(\theta, \phi)/D(\theta, \phi) \qquad (3.8)$$

The alternative ratio that is usually applied to the receive mode is known as the *aperture efficiency* (η):

$$= \frac{\text{Effective area of antenna}}{\text{Physical area of antenna}}$$

Typical practical values for both ratios lie in the range 50% to 75%.

The antenna *equivalent noise temperature* is most important in the receive mode and is related to the *antenna radiation resistance*, which is not a physical value but more an equivalent one. If a transmitter is loaded with a resistor of R ohms and it dissipates the same amount of heat energy as an antenna radiates in the form of electromagnetic energy, then the antenna radiation resistance is R ohms.

This noise component is generally relatively small. A typical antenna noise temperature might lie in the range, 50 K to 100 K. However, the overall noise temperature depends upon the direction in which the antenna is *looking*, as shown in Fig. 3.5. In addition, the noise temperature can be exaggerated by a side lobe that *looks* along a noisy earth.

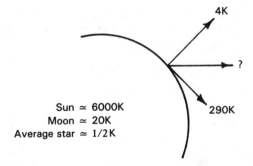

Fig. 3.5 Noise temperature 'seen' by an antenna.

3.4 SECONDARY ANTENNA PROPERTIES

The G/T Ratio

This is one of the system parameters that is very much controlled by the antenna. The factors that affect this gain/temperature (G/T) ratio are the antenna gain (G_a), its equivalent noise contribution (T_a) and the system equivalent noise temperature (T_s). Since G_a is a function of operating frequency and T_a depends on the angle of elevation of the antenna, both should be stated in a system specification. The importance of this *figure of merit* lies in the choice that it presents to the system designer. A particular G/T ratio can be achieved with a large, high-gain antenna, or a smaller antenna combined with a lower noise specification system.

A receiving system has a sensitivity threshold specification, effectively related to the overall gain. This represents the lowest input signal level that will produce just acceptable quality of output or bit error rate.

The ratio is usually expressed in decibels per degree Kelvin and the following equation is commonly used:

$$G/T = 10 \log[\alpha\beta G_a/(\alpha T_a + (1-\alpha)290 + T_s)] dB/K \qquad (3.9)$$

where α represents the coupling losses between the antenna and the system, and β represents the pointing and polarisation losses.

Both coefficients are typially of the order of 0.5 dB. The coupling losses are caused by the small impedance mismatch that occurs between the antenna and block convertor. These are usually shown in the system specification under *voltage standing wave ratio* (VSWR) for the antenna output and block convertor input. The total standing wave ratio is then the product of the two values. It is shown (3) that α can be calculated from the equation:

$$\alpha = \frac{4}{2 + VSWR_t + (1/VSWR_t)} \qquad (3.10)$$

The following figures are extracted from typical systems designed for Ku Band television reception and are presented here to show the relationship between the specification and the G/T ratio.

Antenna gain	48 dB
Pointing and coupling losses	1 dB
Antenna noise temperature	60 K
System noise factor	2.7 dB
Total noise temperature	310 K
Total noise temperature (dBK)	24.9 dBK
G/T ratio	22.1 dB/K

It will be recalled from chapter 2 that the effective equivalent noise temperature (T_e) referred to the system input is given by:

$$T_e = T_a + T_s/G_a$$

and the overall noise factor by:

$$F_o = F_a + (F_s - 1)/G_a$$

where $F_a = 1 + T_a/290$.

Noise factor 2.7 dB $= 1.86 = F_s = 1 + T_s/290$
$$F_s = 1 + T_s/290 = 1.86$$
$$T_s = 250 \text{ K}$$
Total noise temperature $= 250 + 60 = 310$ K, or in dBK
$$= 10 \log 310 = 24.9 \text{ dBK}$$
System G/T ratio $= 48 - 1 - 24.9 = 22.1$ dB/K.

(To convert K into dB, the formula dB $= 10 \log (K/290 + 1)$ can be used.)

The Parabolic Reflector

The mathematical equation for a parabola is:

$$y^2 = 4ax \qquad (3.11)$$

Such a shaped curve, shown in Fig. 3.6, has some special properties. Any ray emanating from point 'a' will reflect off the curve along a path parallel to the x axis. If the curve is rotated around the x axis, the *surface* so produced, is a *parabolic dish*. If this surface is used as a transmitting device, energy emanating from 'a' will be reflected to form a beam parallel to the x axis. Conversely, energy received along a complimentary path, will be concentrated at 'a', the *focal point*.

Another property of the surface is that the total distance, by any path, from the focal point to the *aperture plane* by reflection is constant. Thus as a transmitting device, any energy leaving 'a' will theoretically pass through the aperture plane completely in phase. It is these properties that are responsible for the forward gain of the device.

For a practical dish of diameter D, centre depth C and focal length F, equation 3.11 can be re-written as:

$$
\begin{aligned}
(D/2)^2 &= 4FC \\
D^2 &= 16FC \\
C &= D^2/16F \\
C &= D/(16(F/D))
\end{aligned}
\tag{3.11}
$$

showing that the depth of dish depends on the diameter and the F/D ratio. The dish shape can be completely described from a knowledge of this ratio and either F or D. With the F/D ratio = 0.25 the focal point lies on the aperature plane. For F/D less than this the focal point lies within the dish volume. Whilst this gives a good side-lobe response, the forward gain is lower because the feedhorn fails to *see* the whole surface. For F/D greater than 0.25, *spill-over* can occur and the side-lobe response degrades. In general, a high F/D ratio produces a flatter dish and optimises the forward gain, whilst a lower value gives a better degree of side-lob suppression. For a typical compromise, values between 0.35 to 0.45 are commonly used for prime focus dishes and 0.6 to 0.7 for the offset focus versions.

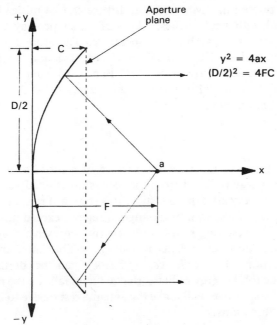

Fig. 3.6 The parabolic reflector.

Accuracy of Parabolic Profile

The overall efficiency depends on the aperture efficiency, phase errors at the aperture plane, *spill-over*, and *blocking losses*. The latter two points will be dealt with in the next section.

The aperture efficiency has been shown as the ratio of effective and physical areas, where the effective area is related to the area of the aperture plane. The efficiency will be greatest when the transmitted energy evenly illuminates the whole surface. It is shown in the literature (4) that because of the small space attentuation between the focal point and the dish surface, the illumination will tend to diminish towards the edge.

This depends on the F/D ratio, and the edge illumination relative to that at the centre is given by:

$$\{1 + (D/4F)^2\}^{-2} \tag{3.12}$$

Substituting typical values for F/D in this equation shows that the edge illumination improves with an increase in F/D.

Phase errors are largely affected by surface inaccuracies, and so are wavelength dependent. As a general rule of thumb, the peak discrepancies should not exceed $\pm$ $\lambda/8$ from the true shape. The problem thus becomes exaggerated as operating frequency rises. Often the surface errors for a dish are quoted in RMS values and these can be misleadingly small when compared with peak values. The surface shape may well become distorted by changes of ambient temperature. Therefore for the higher operating frequencies (Ku Band), solid construction becomes very important. For the lower frequencies (C Band), wire mesh surfaces are sometimes used. Here a useful rule of thumb suggests that the holes should not be greater than about $\lambda/10$.

Spill-Over

The most effective way to utilise the surface of a reflector antenna is to ensure that the energy radiated from the feed is distributed across its area. With simple feeds, any attempt to achieve this state will result in radiating a significant amount of energy into the angular regions exceeding the angle 2θ of Fig. 3.7(a). The resulting *spill-over* of energy represents a signal loss to the forward radiation, as well as radiating possible interference in an unwanted direction. The *spill-over efficiency* can be defined as the percentage of the total energy radiated that actually falls on the reflector. In the receive mode, spill-over will make the antenna responsive to interference from the same directions.

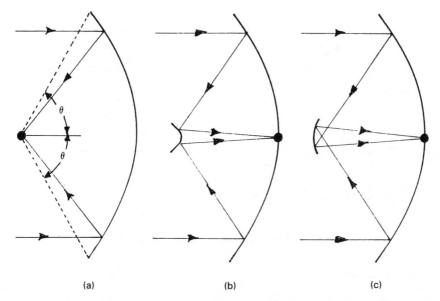

Fig. 3.7 Three methods of feeding the parabolic antenna: **(a)** prime focus; **(b)** Cassegrain sub-reflector; **(c)** Gregorian sub-reflector.
Key: ● = Electronics unit; Θ = Semi angle subtended by parabola.

It is shown in reference (5) that the most effective way of using the source energy is to arrange for an *illumination taper* with the energy level at an angle of 2θ, to be approximately -10 dB relative to that at the reflector centre. It is also shown that this can be achieved using a tapered feed horn, either rectangular or circular, similar to that shown in Fig. 3.8(a). Since the angle 2θ is depdendent upon the dish F/D ratio, there is an optimum relationship between these parameters and the feed horn taper, the design of which has to take into consideration the fact that the radiation amplitudes in the E and H planes might not be symmetric. It is conjectured that this is due to E field fringing which results in the two components of the wavefront *seeing* different values of free space impedance. In the receive mode, this produces different focal points or phase centres for each of the E and H components.

Scalar feed horns of the type shown in Fig. 3.8(b) are commonly used for reception purposes. These have 3 to 7 concentric rings, that behave as quarter-wave choke slots, to correct the E field fringing problem. Such horns are insensitive to polarisation, have good beamwidth and bandwidth, very low VSWR and good side-lobe discrimination. Their use can add an addition gain to the antenna system of about 1 to 2 dB.

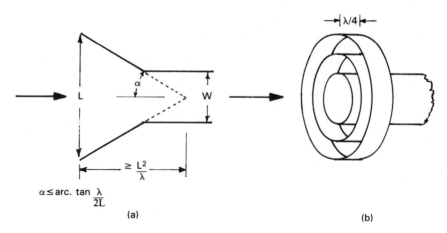

Fig. 3.8 Feedhorns: (a) tapered; (b) scalar.

Blocking Losses

There are two basic methods of illuminating the reflector antenna: either from the focal point using the *prime focus* method depicted in Fig. 3.7(a), or by using a sub-reflector as in Figs. 3.7(b) and (c). The latter diagrams show two alternative sub-reflectors: the convex hyperbolic shape of the *Cassegrain* feed, and the concave elliptical shape of the *Gregorian* feed. All three methods have certain advantages, but in general the sub-reflector methods yield an overall antenna gain of about 75% to 80% compared to about 55% for the prime focus feed system. The accuracy of the sub-reflector surface is most important, otherwise spill-over will be increased.

The prime focus system is the easiest to set up, but with the electronics unit positioned near to the focal point, heat from the sun's rays can produce a problem. Using sub-reflectors places the electronics unit at the back of the dish and in a more convenient position for servicing. Using either method results in physical components being mounted near to the focus. These and their support structure then throw a shadow on the dish surface that reduces the effective area by *blocking*, as shown in Fig. 3.9(a). The structure in front of the dish causes scattering of the electromagnetic energy, which leads to a loss of gain and degradation of side-lobe and cross polar discrimination.

The blocking effect can be completely eliminated by using an *off-set* feed, as shown in Fig. 3.9(b). The main reflector is a section of a larger paraboloid, which can be illuminated by either of the three methods described above. It should be noted that the offset feed produces a bore-sight that is approximately 28° off centre in the elevation plane (6). This is shown in Fig. 3.9(b), together with a representation of the effective area of the reflector surface. In the transmit mode, the offset feed antenna

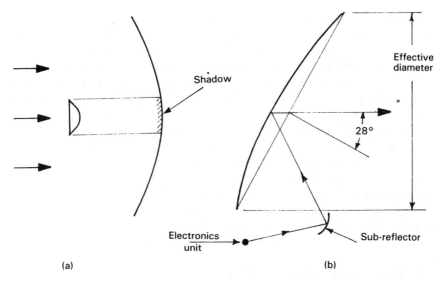

Fig. 3.9 (a) Blocking losses. (b) Offset feed antenna.

introduces a small cross-polar component, which can be corrected by using a suitably shaped sub-reflector. The focal point for either the Cassegrain or Gregorian sub-reflector should coincide with the focal point of the main reflector. The removal of the blocking effect gives increased gain, better side-lobe and cross-polar discrimination and a good VSWR.

Antenna Pointing Losses

The sun and moon, in particular, exert gravitational forces on a geostationary satellite that cause it to drift in a north/south-orientated figure-of-eight path during each sidereal day. By comparison, the drift in the east/west direction is small, but the satellite has to carry fuel for *station keeping* to constrain this drift within acceptable bounds. The larger ground stations with high-gain, narrow-beamwidth antennas need to use servo-controlled tracking systems to maintain good S/N ratios. But when using small fixed receiving antennas, there will be a small daily change in signal level.

It has been shown that the approximate beamwidth (-3dB) is given by $57.3\lambda/D$. Using a 1.8 metre antenna at 10 GHz gives a beamwidth of about 1°, so a pointing error of 0.5° will give rise to a loss in signal level of about 1 dB. Such a calculation also shows the importance of a rigid mounting construction even for a small dish, particularly in windy environments.

G.E.C. McMichael Ltd (Marconi Communication Systems Ltd) produced a series of transportable antennas, one of which was adopted for the Newshawk Satellite News Gathering System. The main reflector is elliptical, with a width-to-height ratio of 2:1. A Gregorian-type sub-reflector provides an offset feed for both receive and transmit modes. The whole construction makes for high efficiency, portability and flexible operation. The width gives high gain and narrow beamwidth in the azimuth plane, to discriminate between adjacent satellites. The reduced height gives a wider beamwidth in the elevation plane to minimise the pointing loss effect due to satellite drift.

The unit consists of a closely matched reflector, sub-reflector and wave guide horn. The surface shapes and illumination taper have been produced by an iterative computer-controlled process known as *diffraction profile synthesis*. As a result, the surfaces are complex and not describable by any simple mathematical function. This design technique produced an antenna with very good side-lobe rejection and cross-polar discrimination, with low noise temperature and high efficiency.

3.5 CHOICE OF ANTENNA TYPE

Satellite communications provide applications for a very wide range of frequencies, extending from about 135 MHz upwards. It has been shown (equation 3.4) that the gain of a parabolic reflector antenna is proportional to its diameter and the operating frequency. Therefore below some frequency its gain must fall below an acceptable level. It is shown (5) that this is likely to be around a diameter of 10λ, below which it might be practicable to change to an alternative type. The Yagi array has been extensively used for terrestrial systems up to about 1 GHz for linear polarised signals, but for circular polarisation its use invokes a -3 dB penalty. Typical maximum gain for a Yagi is around 18 dB. Stacking two similar arrays in parallel yields only a further theoretical gain of 3 dB. Therefore at this stage the Yagi becomes unwieldly, particularly if it is to be steerable. Helical antennas have a useful gain up to about 20 dB for circular polarisation with good directivity. A small commercially available antenna is available for operation in the 1.50 to 1.55 GHz band. This has a figure of merit (G/T) of -23 dB/K down to 5° of elevation. However, as with the Yagi, any further increase in gain tends to be mechanically inconvenient.

There are applications where the antenna gain is of secondary importance to a wide beamwidth. Services using the polar orbiting satellites often need a beamwidth in the order of 180°. The *resonant quadrifilar helix antenna* can meet such a need.

3.6 HELICAL ANTENNAS

The helical beam antenna depicted in Fig. 3.10 is particularly suitable at frequencies below about 500 MHz. The polarisation is circular along the

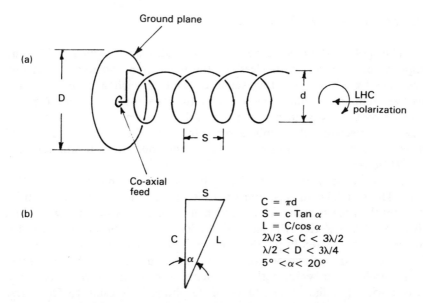

Fig. 3.10 The helical beam antenna and its important dimensions.

helix axis and dependent upon the winding direction. A helix for left-hand circular (LHC) polarisation is shown in Fig 3.10 (a). The dimensions that are important to the characteristics are shown in Fig. 3.10(b). The antenna radiation resistance is low and suitable for use with co-axial cable feeders.

When the circumference of the antenna is approximately one wavelength (λ), the radiation pattern is almost entirely in front of the ground plane. The directivity is empirically given (7) by:

$$D = 15n(S/\lambda)(C/\lambda)^2 \tag{3.13}$$

where:

$C = \pi d$ the circumference
d = diameter
S = $C \tan\alpha$
α = helix angle
n = number of turns
λ = operational wavelength

For the commonly used values of $\alpha = 12°$, $C = \lambda$, the forward gain for n between 3 and 25 turns is given approximately by:

$$G \ dB = 7 + 2\sqrt{n}. \tag{3.14}$$

Additionally, the antenna has a wide bandwidth and very good VSWR. Due to the helical construction, it is physically short in terms of wavelengths.

Resonant Quadrifilar Helix Antennas

Circular polarised antennas with a beamwidth approaching 180°, find applications as receiving devices for the navigation and weather services provided from polar orbiting satellites. Due to such a beamwidth, the receiver remains within a good S/N ratio area for the duration of the satellite pass. Most antennas that meet this requirement for VHF and low UHF are mechanically cumbersome. However, the resonant quadrifilar helix antenna not only has a wide beamwidth but also a useful gain and small size (8).

There is a family of these devices, whose construction is typified by Fig. 3.11. The helix shown is wound for RHC polarisation (reversing the helix reverses the polarisation). The helices consist of thin brass tapes wound on to a dielectric material cylinder, with the lower ends short circuited and the upper ends ·connected in parallel pairs as shown. The total length of each helix is a multiple of two or more quarter wavelengths. That shown has two loops, each of one wavelength total length. The antenna forms a balanced circuit with a radiation resistance of 25 ohms. It is usual to feed these antennas via a sheathed *balanced-to-unbalanced* (balun) transformer through the aperture at the short-circuit end and using co-axial cable. The antennas with three or more quarter wavelength helices have a higher gain, but the half wavelength version has a better front-to-back ratio.

In order to obtain the correct polar diagram with maximum response along the helix axis as shown in Fig. 3.12, it is necessary to feed the two loops in phase quadrature (90°). This can be achieved either by using a 3 dB directional coupler and feeding each loop from separate outputs, or by making one loop resonant at a higher frequency than the other but equally spaced across the operating frequency. By careful control of the reactive components, the current in one loop then leads the applied voltage by 45° whilst the current in the second lags by 45°. The operating bandwidth depends on maintaining this phasing.

Feedback Instruments Ltd UK have produced such an antenna for their Weather Satellite Receiving System, where the two loops are identical except that one contains a pair of loading coils. The two loops are connected in parallel through a balun to give a feed impedance of 25 ohms. Their design includes a built-in high-gain pre-amplifier and a five-section low-pass filter, which avoids the need to match the 25 ohms to a 50 ohms co-axial cable. This removes one source of signal loss, the amplifier power being provided over the feeder cable. The complete antenna is coated with glass fibre to give protection from the elements.

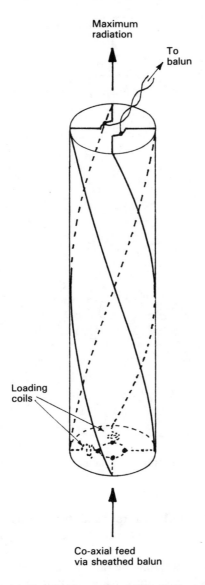

Maximum
radiation

To
balun

Loading
coils

Co-axial feed
via sheathed balun

Fig. 3.11 A half-turn quadrifilar helix antenna.

3.7 REFLECTION, REFRACTION, DEPOLARISATION AND FEEDS

The behaviour of light waves incident upon a boundary layer between two media, is described by two laws:

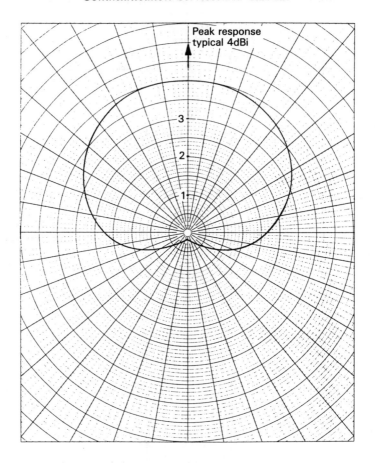

Fig. 3.12 Quadrifilar helix antenna polar diagram.

(1) The incident ray, reflected ray and the normal-to-the separation surface between the two media lie in the same plane at the point of incidence.

(2) Snell's law, which states that the ratio of the sines of the incidence and refraction angles is constant for a given pair of media. When one of the media is a vacuum (or air to a close approximation), this ratio gives the *refractive index* n for the second medium. The refractive index is also the ratio of the velocity of light in free space to its velocity in the medium.

Using Fig. 3.13(a), Snell's law can be expressed as:

$$\sin \theta_3 / \sin \theta_1 = n_1/n_2 = \text{constant} \qquad (3.15)$$

where:

θ_1 = angle of incidence
$\theta_2 = \theta_1$ = angle of reflection
θ_3 = angle of refraction
n_1, n_2 = optical refractive indices for the two media

It will be recalled from electromagnetic wave theory that the velocity of light c in free space is given by:

$$c = 1/\sqrt{(\mu_0 \epsilon_0)}$$

where μ_0 and ϵ_0 are the absolute values of permeability and permittivity for free space respectively.

The behaviour of electromagnetic waves (which, like light waves, are part of the same electromagnetic radiation spectrum) incident upon a surface, is a fundamental feature of several antenna types.

For electromagnetic waves, equation 3.15 can be rewritten as:

$$\sin \theta_3 / \sin \theta_1 = \sqrt{(\mu_1 \epsilon_1 / \mu_2 \epsilon_2)} \qquad (3.16)$$

where μ_1, μ_2, ϵ_1 and ϵ_2 are the relative permeabilities and permittivities of the two media respectively. If one of the media is air and the other a dielectric material, μ is approximately equal to μ_0 so that equation 3.16 simplifies to:

$$\sin \theta_3 / \sin \theta_1 = \sqrt{(\epsilon_0 / \epsilon_2)} \qquad (3.17)$$

Since $\epsilon_2 < \epsilon_0$, the velocity of propagation and the wavelength in the dielectric will be less than in air. From equations 3.15 and 3.17, it can be shown that the proportional reduction of velocity and wavelength is dependent upon $\sqrt{\epsilon_r}$, where ϵ_r is the relative permeability of the second medium.

If a wave is incident upon a perfect conducting surface, the angle of reflection will be equal to the angle of incidence but the wave polarisation will have been changed. If the incident wave is right-hand circular polarised (RHCP), the reflected wave will be left-hand circular polarised (LHCP) and vice versa, as shown in Fig. 3.13(b).

Fig 3.13(c) shows the wave incident upon a dielectric material, with the H field perpendicular to the boundary layer. A critical angle (Brewster angle) can be found, where the wave is totally transmitted into medium 2. If the incident wave were circular polarised (CP) or elliptically polarised (EP), the transmitted wave would be elliptically or linearly polarised respectively.

All four states of polarisation are used for different satellite services, the state chosen for a particular application depending upon some particular attribute of that state. For instance, *Faraday rotation* within the ionosphere, the change of satellite attitude or the misalignment of an antenna feed can each give rise to a change of plane of polarisation, with the attendant loss of received signal. Therefore circular polarisation, which is

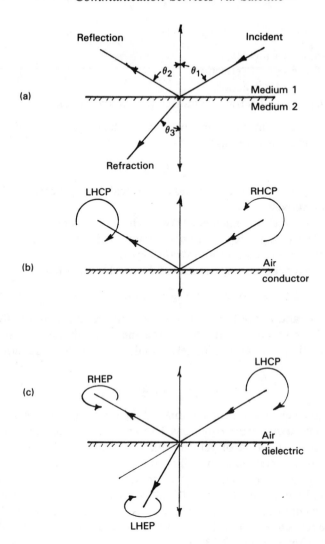

Fig. 3.13 (a) Reflection and refraction. **(b)** Reflection from a conductor surface. **(c)** Refraction in a dielectric material.

not affected in this way, would be used. Both opposite linear and circular polarisations give rise to similar degrees of isolation under frequency re-use conditions. A linear feed will receive both RHCP and LHCP equally well, but will only respond to one component in the wavefront, resulting in a 3 dB loss of possible signal level.

Dielectric Depolariser for CP Feeds

The dielectric depolariser can be used to modify circular polarised waves so that they may be received by a linear feed without significant loss of signal. This can be done by reducing the value of δ (Fig. 3.2(b)) to zero. To understand how this can be achieved, some fundamental features of wave motion are now re-examined.

The Phase Velocity v_p of a wave $= \lambda/T = \lambda.f$ where λ, T and f have their usual meanings. If the phase velocity per unit length is β, then over a wavelength λ the phase shift is 2π radians, so that:

$$\beta = 2\pi/\lambda \text{ or } \lambda = 2\pi/\beta$$

Now $v_p = \lambda.f = 2\pi f/\beta = \omega/\beta$, where ω is the angular velocity in rd/sec.

Consider the case of a thin slab of dielectric as shown in Fig. 3.14(a), which is placed in a waveguide through which a wavefront is flowing The component E_y will still propagate largely through air, whilst the component E_x must propagate through the dielectric, but with a shorter wavelength, so that its phase shift is represented by:

$$\beta y = 2\pi/\lambda_y \text{ and also } \beta_x = 2\pi/\lambda_x$$

For the dielectric to produce a phase difference of $\pi/2$ rd. (90°) between E_x and E_y, its length L must be such that:

$$\beta_y L - \beta_x L = \pi/2 \tag{3.18}$$

Then the slab effectively converts circular into linear polarisation. The slab must be mounted in the waveguide/feedhorn at an angle of $\pm$ 45° to the feed probe as shown by Fig. 3.14(b). In a practical feed system, it is possible to arrange for the depolariser to be rotated through 90° to change from RHCP to LHCP, and vice versa.

Feeds for Linear Polarisation

To receive signals of both x and y linear polarisations usually requires the use of two low-noise block-converters (LNBs). These may be mounted in series along a circular waveguide as shown in Figs. 3.15(a) and (b). Each LNB has its own feed probe and the two are at right angles to each other. Rotating the assembly to maximise the signal level for one polarisation automatically assures maximum signal at the output of the alternate LNB, selection of the necessary LNB being achieved by switching dc power supplies.

Alternatively, the two LNBs can be fed in parallel via an *ortho-mode-transducer* (OMT) in the manner indicated in Figs. 3.16(a) and (b). The OMT consists of a circular section of waveguide with a rectangular branch

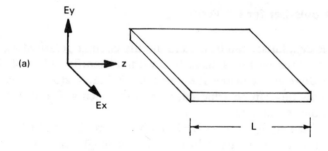

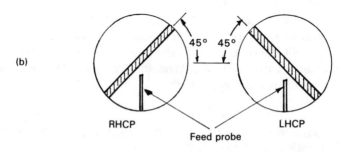

Fig. 3.14 (a) The dielectric depolariser. **(b)** View of depolariser looking into a waveguide/feed horn.

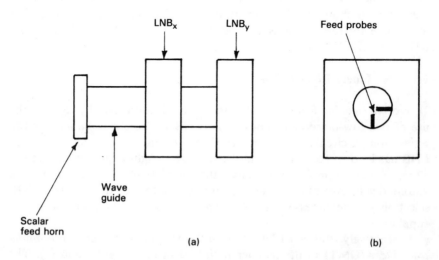

Fig. 3.15 Dual linear feeds in series.

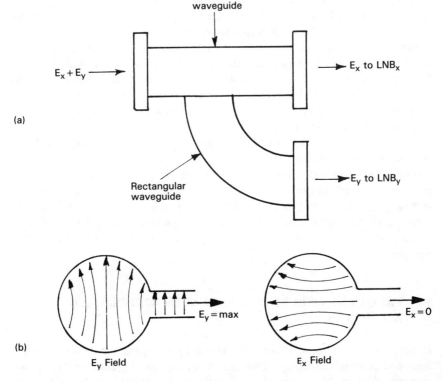

Fig. 3.16 (a) The orthomode transducer; **(b)** Coupling of E field into rectangular section.

section. By rotating the OMT, the linear polarised E_y field will couple with the narrow dimension of the rectangular waveguide to transfer its signal to LNB_y, leaving the E_x component to continue straight through to LNB_x. The rectangular opening thus behaves as a low impedance to one component and a very high impedance to the other one.

A further alternative that avoids the use of a second LNB involves rotating the feed probe within the waveguide by 90° using an electro-mechanical polariser driven by a stepper or servo motor. Whilst this gives a good cross-polar response, the mechanical parts tend to wear, or seize up due to weather conditions.

Faraday Rotation

An electromagnetic wave is influenced by the earth's magnetic field as it propagates through the ionosphere, in such a way that causes a rotation of the plane of polarisation of a linear polarised wave. This is known as

Faraday rotation, the degree of rotation depending on the ionospheric path length and the intensity of the magnetic field. This phenomenon forms the basis of operation of several microwave devices.

Faraday rotation is increased if the wave is constrained within a waveguide, with a dc magnetic field applied externally. The effect can be further enhanced if a relatively small piece of ferrite is judiciously placed in the waveguide. Ferrite has the highly desirable magnetic properties of high permeability and high internal molecular resistance with very low loss, even when used at microwave frequencies.

By a suitable choice of ferrite dimensions and strength of magnetic field, the plane of polarisation can be rotated through 90° over a fairly short distance. Alternatively, the external magnetic field can be switched, to provide ± 45° of rotation with about ± 40 mA of current. The degree of loss with such a dual polarity feed is typically less than 0.3 dB. A cross-polar attenuation of about 20 dB is typically achieved with such a device, and a *fine* or *skew* adjustment is often provided to maximise the performance.

3.8 FREQUENCY-SELECTIVE SURFACES AND PLANAR ANTENNAS

In certain applications, it may be necessary to operate on two different frequencies simultaneously. This could, of course, be achieved with two antenna systems. However, if a sub-reflector is constructed from dichroic material, which is transparent at the prime focus feed frequency and highly reflective at the frequency provided from a Cassegrain/Gregorian feed mounted in the conventional position, the cost of an antenna system can be saved. *Frequency-selective surfaces* have such properties.

These are arrays of metallic elements printed over the surface of a suitably shaped dielectric substrate, using normal printed circuit techniques. The surfaces behave as high-pass filters, transparent to the prime focus feed whilst being highly reflective to the lower frequencies. The elements have a self-resonant frequency that depends upon their dimensions and is responsible for these characteristics. Much work has been reported (9) using arrays of elements of various shapes, ranging from the *Jerusalem cross* of Fig. 3.17(a) through concentric rings and squares, to the *tripoles* shown in Fig. 3.17(b). Very low losses, ranging from 0.1 to 0.5 dB have been reported for both transmissive and reflective modes, with good cross polar performance for linear polarised signals.

In the reflective mode, the elemental microwave patches of the frequency-selective surface radiate as antennas that have been fed from the incident energy of the electromagnetic waves. If the patches on such a surface are coupled together and fed from a source such as a transmitter, then the surface will radiate energy in a similar way.

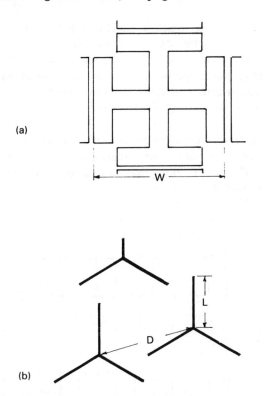

Fig. 3.17 Frequency selective surface patterns: **(a)** Jerusalem cross; **(b)** tripole.

Although radiation emanates from the satellite antenna as a spherical wave, when received at the earth station it has become a very close approximation to a plane wave. Under such conditions, any flat panel carrying an array of resonant circuits will collect radiated energy that lies within some particular bandwidth. The *printed planar antenna* is an array of microstrip radiating elements fed by microstrip transmission lines, the construction being similar to double-sided printed circuit boards. The patterns for the elements and feed lines are formed by etching away copper cladding from one side of the substrate, leaving the other side to act as a ground plane. The substrate used has varied from high-grade glass fibre to polythene, but needs to be chosen for low loss with adequate mechanical stiffness.

Over the development period, many different patch shapes have been studied, showing that both linear and circular polarisations can be radiated or received. Up to 1986, gains of up to 35 dBi and operating frequencies up to 85 GHz had been reported (10, 11), suggesting that the concept might provide a valuable alternative antenna for mobile satellite communications. Because of their flat nature, they adapt well to use with ground vehicles or aircraft. In these cases, a vehicle's metallic skins can act as the ground plane giving rise to the term *conformal antennas*.

The chief advantages of planar arrays lies in the thin profile. This makes them less obtrusive and adaptable as comformal antennas, with acceptable side-lobe rejection. Further, if these units are fed from a corner, then it becomes very easy to connect four in parallel to improve the overall performance.

Of the many different configurations that have been used, the patterns shown in Fig. 3.18 are popular. The antennas are either made as a complete panel (Fig. 3.18(a)) or in sections that can be coupled together to form a large array (Fig. 3.18(b)).

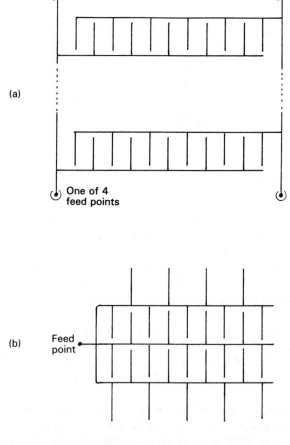

Fig. 3.18 Microwave planar antenna arrays: **(a)** end sections of square array with steerable directivity; **(b)** single section of multiple array.

In terms of directivity, if the panel is fed from the centre then the major part of the radiation will be normal to the ground plane. However, the antenna of Fig. 3.18(a) can be fed from either of the four corners, and this gives rise to four different off-centre beams (11). An alternative way of making the array *squint* is to insert phase shifters in the feeds between panel sections. By carefully controlling the reactive components, the beam may then be directed in a preferred direction (See Fig. 3.19).

For receiving circular polarised waves, a dielectric depolariser can be placed in front of the array, with the dielectric slats at 45° to the elements. This functions in the manner of the depolariser previously described and gives rise to a 3 dB signal penalty.

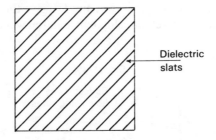

Dielectric slats

Fig. 3.19 Depolariser for planar antenna array.

British Satellite Broadcasting (BSB) now BSkyB, proposed the use of a *squarial* for the DBS service provided from a high-power satellite (approximately 60 dBW). This 40 or 50 cm diamond-shaped square unit provides enough gain to produce excellent picture quality even when working with a C/N ratio as low as 11 dB. The basic structure is designed around either 144 or 256 resonant cavities separated by 0.9λ, each with separate probes. These are arranged to feed an embedded low noise block-convertor (LNB) via a combining network, which introduces some attenuation. This has to be countered by using a dielectric material with very low loss.

3.9 FRESNEL ZONE PLATE ANTENNAS (15)

This antenna is based on the Fresnel principle of a series of concentric rings, alternately transparent and opaque and deposited on a transparent surface. This has the ability to focus electromagnetic waves on to a point behind the surface. The alternate zones are formed by screen printing on to a clear plastic layer or flat glass sheet using reflective or absorbing inks. Silver or graphite loaded ink is used for the plastic surface and metallic oxides can be used on glass. This last form leaves the structure completely transparent to visible light. If the dielectric layer is backed by a metallic surface acting as a reflecting ground plane, the focal point switches to being a mirror image of the transmissive version. This has the effect of increasing the gain of the reflector by almost 3 dB. The principle of both modes is shown in Fig. 3.20. Furthermore, if the rings are made elliptical, the focal point shifts so that electromagnetic waves approaching from a suitable angle *see* concentric circles. The secant of this angle is proportional to the ratio

$$\frac{\text{semi-major axis}}{\text{semi-minor axis}}$$

of the ellipsoid. This feature is therefore used to give the reflector surface a *squint*. Fig. 3.21 shows the relationship between the wavelength dependent dimensions.

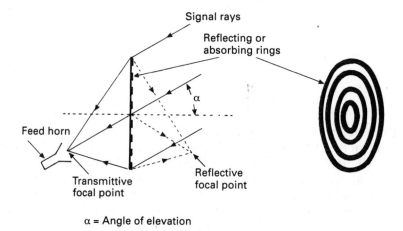

α = Angle of elevation

Fig. 3.20 The elliptical zone plate.

The gain of the reflector is approximately proportional to the number of rings and the beamwidth is inversely proportional to the area. The bandwidth is practically equal to $2\omega_o/N$, where ω_o represents the mid-band frequency and N is the number of rings. For a typical reflector, the bandwidth is about $15\%\omega_o$. Working at 11 GHz, a 90 cm diameter zone plate will give good television reception from a signal source of about 50 dBW. Comparative gain values in dBi are shown in Table 3.2.

3.10   SIGNAL STRENGTH CONTOUR MAPS OR FOOTPRINTS

The contour lines of the satellite's transponder footprint, as depicted by Fig. 3.22 (12), represents the level of signal receivable by a ground station.

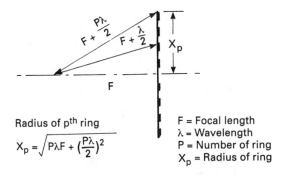

Radius of pth ring

$$X_p = \sqrt{P\lambda F + \left(\frac{P\lambda}{2}\right)^2}$$

F = Focal length
λ = Wavelength
P = Number of ring
X_p = Radius of ring

Fig. 3.21 Signal focusing and dimensions of concentric rings.

Table 3.2

Number of zones	Theoretical gain	Practical gain
1	6.0	4.5
2	11.8	9.0
3	15.2	11.5

Unlike the maps used for terrestrial transmission systems, which are usually scaled in volts/metre (dBV) or microvolts/metre (dBμ), the satellite's contours are scaled either in terms of power flux density (PFD) or effective isotropically radiated power (EIRP). PFD represents the signal level received on earth per square metre of surface (W/m²) or decibels relative to 1 watt/m² (dBW/m²). As an example, if an antenna of effective area of 1m² is exposed to a PFD of − 120 dBW/m², it will collect − 120 dBW of power.

$$-120\ dBW = 10^{-12}\ W = 1pW.$$

Since power is given by V^2/R and R in this case is the free space impedance of 120π ohms. $V = 19.42\ \mu V$ or 25.76 dBμ.

EIRP is related to the transmission polar diagram, and is the product of the signal power fed to the antenna and the transmitting antenna gain in a preferred direction. If the power transmitted from an isotropic source is P_t watts, the PFD at the surface of a sphere will be $P_t/4\pi r^2$ W/m², where r is the spherical radius.

If the isotropic source is now replaced with an antenna of gain G_t, then assuming the same power input, the PFD at the surface will be $G_t P_t/4\pi r^2$ W/m² in some direction. If this energy is incident upon a receiving antenna ·with an effective area of A_e m², the total received power P_r will be:

$$G_t P_t A_e/4\pi r^2\ \text{watts} \tag{3.19}$$

From equation 3.3, $A_e = G_r\lambda^2/4\pi$, where G_r is the gain of the receiving antenna, and making a substitution in equation 3.19, gives:

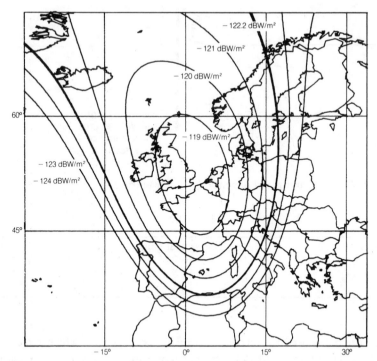

Fig. 3.22 Spot West power flux density contours for ECS-1, F1. (By courtesy of European Space Agency (ECS).)

$$P_r = G_t P_t G_r \lambda^2/(4\pi r)^2 \text{ watts}$$
$$= G_t P_t G_r (\lambda/4\pi r)^2 \text{ watts and}$$
$$P_r/G_r = P_t G_t (\lambda/4\pi r)^2 \text{ watts} \qquad (3.20)$$

Now $P_r/G_r = \text{PFD}$, $P_t G_t = \text{EIRP}$, and the term $(\lambda/4\pi r)^2$ is usually referred to as the *free space attenuation*. Thus the equivalence between the two methods of scaling the footprint contours can be summarised as:

$$\text{PFD dBW} = \text{EIRP dB} - \text{free space attenuation dB} \qquad (3.21)$$

For a geostationary satellite receiving station for 12 GHz, the minimum value of r is 35,775 km. Therefore at this frequency the minimum free space attenuation is approximately 195.1 dB.

3.11 ANTENNA POINTING AND MOUNTING

In addition to the adjustment for focus, the ground station antenna needs to be mounted with two degrees of freedom of movement. This allows the correct angles of *elevation* (El) to the tangent plane on which the antenna is

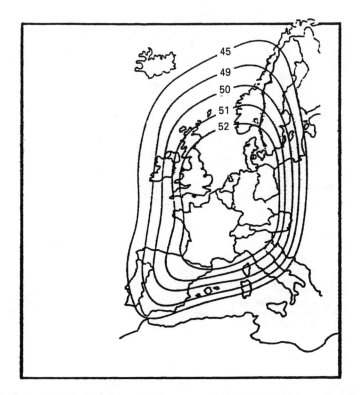

Fig. 3.23 An ASTRA 1B footprint (courtesy of Societé Européene des Satellites (SES)).

assumed to be located, and *azimuth* (Az) or bearing. These two angles are shown as α and Φ in Fig. 3.24, and are dependent upon the latitude and longitude differences between the satellite's orbital location and the earth station's position, shown as S and P respectively in Fig. 3.24(a). Point S_1 is the sub-satellite point on the equator and the line P,S_1 is a 'Great Circle' path. The angle of elevation α is the angle between the line-of-sight path to the satellite P,S and the tangent plane at P along the line P,S_1. The azimuth angle Φ in Fig. 3.24(b) is the angle between P,S and the true north (relative to the fixed stars).

The angle of elevation can be obtained (13) from:

$$\alpha = \text{arc. } \tan((\cos \gamma - 0.151269)/\sin\gamma) \tag{3.22}$$

where:

$$\gamma = \text{arc. } \cos(\cos \beta \times \cos \theta) \tag{3.23}$$

For latitudes above 81°, this value of γ is negative because the satellite is below the horizon.

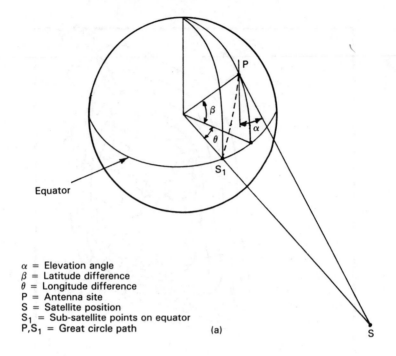

α = Elevation angle
β = Latitude difference
θ = Longitude difference
P = Antenna site
S = Satellite position
S_1 = Sub-satellite points on equator
P,S_1 = Great circle path

(a)

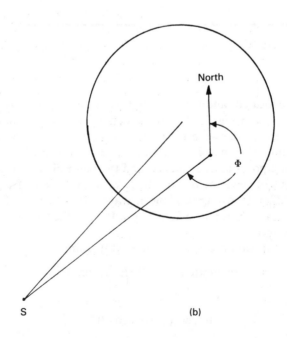

(b)

Fig. 3.24 (a) and **(b)** The pointing angles (look angles).

The angle of azimuth is given by:

$$\Phi = \text{arc. } \tan(\tan \theta / \sin \beta) \tag{3.24}$$

(Adding 180° if the satellite is west of the Greenwich meridian.)

The path length d needed to calculate the free space attenuation can be found from:

$$d \text{ Km} = 35765 \sqrt{1 + 0.41999(1 - \cos \gamma)} \tag{3.25}$$

The mathematical proof of these equations can be found in reference (14). A pocket calculator approach is provided in Appendix A3.3.

As an alternative, the antenna can be provided with a *polar mount*, the principle of which is sketched in Fig. 3.25(a). In this case, it is theoretically only necessary to calculate the elevation angle α and adjust this by positioning the king post at an angle of $(90° - \alpha)$ with the antenna pointing due south. Swinging the antenna around the king post bearings then allows the antenna to sweep out an equatorial arc. In practice, an additional fine adjustment is provided (declination adjustment) to account for the system vagaries. The proof and method for the calculation of α in the polar mount case is given in Appendix A3.4/5, where it is shown that a good approximation for α can be obtained from:

$$\alpha = 90° - 1.15385 \text{ (station latitude°)} \tag{3.26}$$

Since the antenna beamwidths are small, and the dish provides a significant wind surface, the nature of the mount must take this into account. Wind vibrations of about 0.5° or so can cause a fluctuation in received S/N ratio. Thus the relative merits of the two mounting methods tend to depend on antenna size. For a larger antenna, the Az/El method using an A frame approach is more stable. For antennas with a diameter of less than about 2 metres, the polar mount is useful. This has a significant advantage if reception from more than one satellite is intended, as it only requires one servo drive system to track all equatorial orbits.

In those conditions where a greater pointing accuracy is needed, the *step-track* control technique may be used. Each of the azimuth and elevation servo controls cyclically drive the antenna pointing angle in turn, in a series of small steps whilst the received signal level is monitored. (Usually the satellite beacon signal). If a decrease in level is detected then the antenna was moved in the wrong direction and the control system drive is reversed. If an increase is detected, then the antenna is driven a further step in the same direction. To offset the problems of system backlash and propogation variation, a threshold level is set so that the system is inoperative unless the signal level falls below some predetermined value.

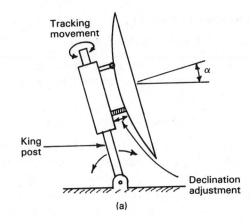

(a)

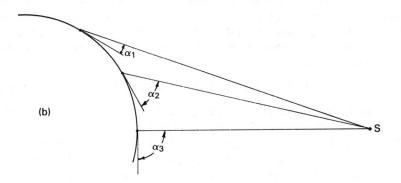

(b)

Fig. 3.25 The polar mount.

REFERENCES

(1) 'Cathode Ray'. *Wireless World*, Oct. 1975. Radio Waves — What makes them go. Pages 469–72.

(2) Slater, J.N. and Trinogga, L.A. 1985 *Satellite Broadcasting Systems: Planning and Design.* Chichester: Ellis Horwood Ltd.

(3) Rijssemus, M., Tratec Bv, Veenendaal, Holland. Private communication to author.

(4) Rudge, A.W. *et al.* (1982) *Handbook of Antenna Design, Vol. 1* London: Peter Peregrinus Ltd.

(5) Christieson, M.L. 'Parabolic Antenna Design' *Wireless World*, Oct. 1982.

(6) Butcher, M.E., British Telecom Research Laboratories, Martlesham Heath, U.K. Private communication to author.

(7) Kraus, J.D. 1984 *Electromagnetics*. 3rd ed. New York: McGraw Hill Book Co.

(8) Christieson, M.L., Feedback Instruments Ltd, Crowborough, U.K. Private communication to author.

(9) Parker, E.A. and Langley, R.J. Antenna Group, University of Kent at Canterbury, U.K. Private communication to author.

(10) Lafferty, A. and Stott, J.H. BBC U.K. Engineering Information Department. Private communication to author.

(11) Mullard Ltd U.K. (1982) *Mullard Technical Publication M81-0147*.

(12) European Space Agency (1983). *ECS Data Book Esa BR-08*.

(13) Wise, F. 1979 'Fundamentals of Satellite Broadcasting'. *IBA (UK) Technical Review No. 11*.

(14) Rainger, P. *et al*. 1985 *Satellite Broadcasting*. London: John Wiley & Sons.

(15) Wright, W. Mawzones Ltd, Baldcock, UK. Private communication to author.

Chapter 4

Microwave Circuit Elements

4.1 TRANSMISSION LINES

When electromagnetic energy flows along a transmission line, the voltage and current distribution along the line depends on the nature of the load at the far end. If the load impedance correctly matches the characteristic impedance of the line, then all the energy travelling on the line will be absorbed by the load. Under mismatch conditions, the load can only absorb an amount of power that is dictated by Ohm's law, any surplus energy being reflected back along the line towards the generator. The forward and reflected waves combine to form *standing waves* all along the line that are indicative of the degree of mismatch. Figure 4.1 shows the voltage and current distribution near to the load for two extreme cases of mismatch, where the load is either an open or a short circuit. In either case, there will be no power absorbed in the load, because either the current or the voltage is zero. Hence there will be a total reflection of energy.

The standing wave pattern depends on the wavelength/frequency of the transmitted signal and is repetitive every half wavelength, as shown by Fig. 4.1(a). The impedance (V/I ratio) seen by the signal thus varies all along the line, being purely resistive of very high or very low value at the $\lambda/4$ points and either capacitive or inductive reactive in between. An open circuit at the end behaves as a very low resistance just $\lambda/4$ away, whilst the short-circuit case behaves in the opposite way.

This impedance transformation property, allows transmission lines to be supported on metallic $\lambda/4$ stubs, as shown by Fig. 4.1(b), without affecting the signal power flow in any way. If the signal frequency is reduced, the stub becomes less than $\lambda/4$ long at the new frequency and so behaves as an inductive reactance to give rise to low-frequency losses. In a similar way, an increase of frequency causes the stub to develop capacitive reactance to limit the high frequencies.

If an infinity of $\lambda/4$ stubs are connected in parallel all along the lines, a rectangular box shape develops which will continue to carry energy. Such a structure is described as a *waveguide*.

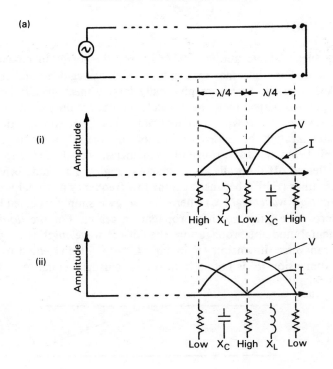

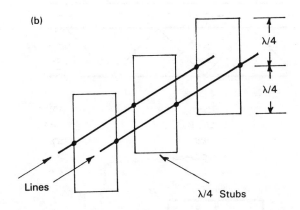

Fig. 4.1 (a) Standing waves on (i) open circuit line, (ii) short circuit line. **(b)** Transmission lines supported by λ/4 stubs.

4.2 WAVEGUIDES

At frequencies above about 1 GHz or so, the losses in co-axial cable transmission lines become unacceptable and waveguides are commonly used. Although the latter are physically larger, mechanically stiffer and more expensive than cables, these disadvantages are outwayed by the very low losses at microwave frequencies. The electromagnetic energy propagates through the guide by reflections off the side walls. Figure 4.2(a) shows this behaviour for a range of frequencies, with the lower frequencies being reflected off the walls at the sharpest angles. The guide behaves as a high pass filter (HPF), because at some low frequency a critical wavelength occurs (*cut-off wavelength* λ_c) where the energy is simply reflected back and forth across the guide so that propagation ceases. For the *dominant* or fundamental mode of propagation, the cut-off wavelength is proportional to the transverse dimension 'a' in Fig. 4.2(b), λ_c being equal to 2a. The narrow dimension 'b' is not so critical and is usually equal to a/2.

(a)

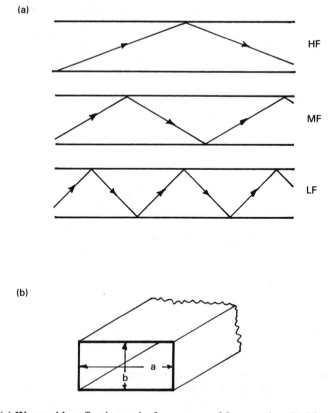

Fig. 4.2 (a) Waveguide reflective paths for a range of frequencies. (b) Dimensions of rectangular waveguide.

The velocity of propagation within waveguides is similar to that in co-axial cables. In the latter case, the wave travel is slowed by the charging and discharging of the cable self-capacitance by a current flowing against its self-inductance. In the waveguide, the energy has to travel via the rather longer reflective path.

For sinusoidal signals, the dominant mode of propagation gives the longest value of critical wavelength and so determines the lower cut-off frequency. Figure 4.3 shows the distribution of the E and H field force lines. The E field lines terminate on the side walls and the H field lines form complete current loops, both propagating through the guide. The lengths of the E field force lines in Figs. 4.3 (a) and (b) represent the sinusoidal variation of field intensity.

(a)

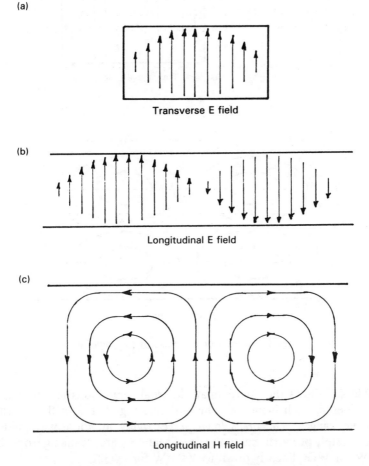

Transverse E field

(b)

Longitudinal E field

(c)

Longitudinal H field

Fig. 4.3 E and H fields for dominant mode propagation.

Higher-order modes can propagate as indicated by Fig. 4.4. In fact, the larger the waveguide (in terms of wavelengths), the greater the number of modes that can propagate. However, the presence of very many modes all travelling at different velocities and following different reflective paths is clearly undesirable. It is therefore usual to chose the waveguide dimensions large enough to support only the dominant mode. This restricts the use of a given rectangular waveguide to a relatively narrow range of frequencies.

Similar modes of propagation occur in circular waveguides, which tend to have a wider bandwidth. However, these introduce mechanical and electrical problems if they are long and include bends.

(a)

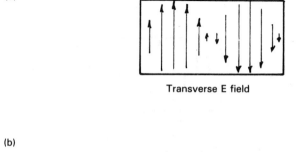

Transverse E field

(b)

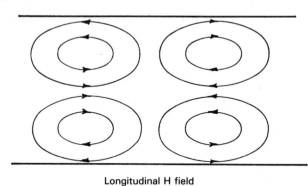

Longitudinal H field

Fig. 4.4 Second-order mode of propagation.

Table 4.1 shows the approximate internal dimensions for the popular waveguides in both inches and millimetres, together with the operating frequency ranges. The *breakdown power rating* for these waveguides is approximately proportional to the cross-sectional area, ranging from about 50 mW for WG6 linearly down to 100 μW for WG22.

Table 4.1

Type number	Internal dimensions (a,b, in (mm))	Frequency range (Dominant mode GHz)
WG6	6.5, 3.25 (16.5, 8.25)	1.12– 1.7
WG8	4.3, 2.15 (10.9, 5.46)	1.7 – 2.6
WG10	2.84, 1.34 (7.2, 3.4)	2.6 – 3.95
WG12	1.87, 0.87 (4.75,2.22)	3.95– 5.85
WG14	1.37, 0.62 (3.49,1.58)	5.85– 8.2
WG16	0.9, 0.4 (2.29,1.02)	8.2 – 12.4
WG18	0.62, 0.31 (1.58,0.79)	12.4 – 18
WG20	0.42, 0.17 (1.07,0.43)	18 – 26.5
WG22	0.28, 0.14 (0.71,0.36)	26.5 – 40

4.3 RESONANT CAVITIES

At lower frequencies, a parallel combination of inductance (L) and capacitance (C) provides a resonant circuit across which maximum voltage will be developed at some frequency. Progressive reduction of both L and C will cause the resonant frequency to increase. Figure 4.5 shows a limiting condition where C has been reduced to a pair of parallel plates, whilst L has been reduced by connecting the plates together with parallel straps that behave as inductors. The ultimate limit is reached where the two plates are connected together with the four sides of a box. This structure is a *resonant*

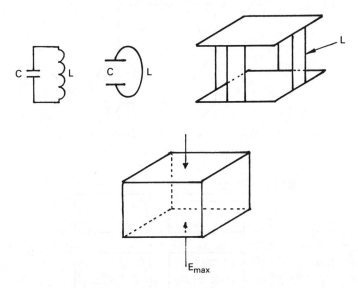

Fig. 4.5 Development of resonant cavity.

cavity, and if energised at an appropriate frequency will develop maximum voltage across the points shown.

Figure 4.6(a) shows that this structure has a resonant wavelength dictated by the length of its diagonal, the electromagnetic field reflecting off the sides as indicated. Figure 4.6(b) shows how the fields oscillate at quarter-period intervals when the cavity is energised, the total energy being transferred between the electric and magnetic fields in an oscillatory manner.

The Q factor, which is very high, is typically about 15000 (unloaded) and is given by:

$$Q = 2\pi(\text{Total energy stored in cavity/Energy lost in 1 cycle due to resistivity of walls}) \qquad (4.1)$$

From which it can easily be deduced that Q is proportional to the ratio of cavity volume to interior surface area.

4.4 COUPLING TO CAVITIES AND WAVEGUIDES

Of the various ways in which signals may be coupled into cavities and waveguides, those shown in Fig. 4.7 are common. The probe in Fig. 4.7 (a) acts as a dipole aerial generating or responding to the E field. The current

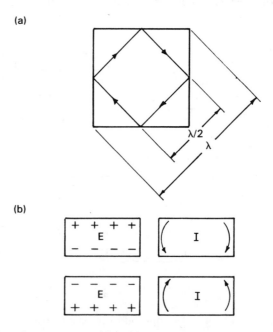

Fig. 4.6 Oscillations in resonant cavities: **(a)** resonant wavelength; **(b)** oscillating E and H fields.

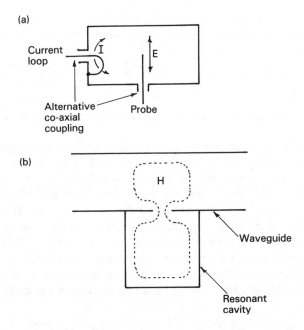

Fig. 4.7 Coupling to cavities and waveguides.

loop either generates or responds to the H field. Figure 4.7 (b) shows how a resonant cavity can be coupled to a waveguide via a hole or *iris*, coupling being achieved through the H field. Cavities of different resonant wavelengths may be coupled to a waveguide in this way, to form a filter circuit. Cavities can be retuned by introducing a metallic screw or dielectric rod into the cavity, from the centre of either of the large area sides. In each case, either the inductance or capacitance of the cavity is increased, to lower the resonant frequency.

4.5 CIRCULATORS AND ISOLATORS

Two ferrite devices that are used extensively in microwave circuits are circulators and isolators. The most commonly used circulator, shown schematically in Fig. 4.8(a), has three ports each spaced by 120°. When each port is correctly terminated, a signal input at one port appears as an output at the next, the third port being isolated from the signal. The behaviour at each port changes in a cyclic manner as indicated in Table 4.2.

Such a device is very suitable for separating incident and reflected signals. The insertion loss in the transmissive direction is typically less than 0.5 dB, whilst the attenuation at the isolated port is better than 20 dB. As an example, port 1 might be driven from a transmitter, whilst port 2 provides the antenna feed. In the receive mode, port 2 becomes the input and the

Table 4.2

	Ports	
Input	Output	Isolated
1	2	3
2	3	1
3	1	2

circulator passes this signal to port 3, providing an isolation between transmitter and receiver of better than 20 dB. For this high-power application, the circulator would be positioned within the waveguide feed to the antenna. The basic construction of a low-power device is shown in Fig. 4.8(b). A small, thin slab of ferrite acts as a substrate for a thin film-deposited gold electrode. A permanently magnetised bias magnet sits on top of the electrode structure, to generate Faraday rotation within. Inputs and outputs are provided by *microwave integrated circuits* (MIC).

The behaviour of the device depends upon the combined effects of the magnetic fields due to the input signal power and that provided by the bias

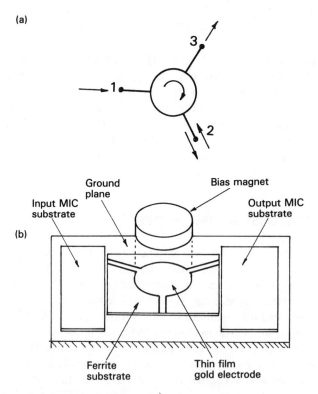

Fig. 4.8 (a) Schematic diagram of circulator. **(b)** Circulator construction.

magnet. Figure 4.9(a) shows the lines of the electric and magnetic fields within the substrate when the bias magnet is not in place. Ports 2 and 3 are at the same potential and are effectively isolated from port 1. By applying a correctly chosen bias level from the permanent magnet, the field components can be rotated by 30°. Port 3 is now at a null in the electric field, whilst ports 1 and 2 are at equal but opposite potentials. Conduction can occur between ports 1 and 2 but port 3 will be isolated from the signal. By reversing the bias field, the signal circulation would be reversed.

An isolator is a two-port device that functions in a similar way, with similar levels of insertion loss and attenuation. In fact, a circulator can behave as an isolator if the third port is terminated in a resistance equal to its characteristic impedance. Used in this way, the reflected energy is absorbed in the resistive load. Such a device is ideal to use in the coupling path between an RF oscillator and its load, to prevent frequency pulling.

4.6 DIELECTRIC RESONATORS

These ceramic devices, made as small discs of sintered *barium titanate*, are used at microwave frequencies in the same way as quartz crystals are used at

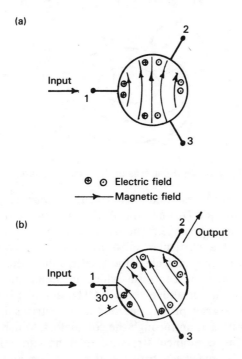

Fig. 4.9 (a) Field pattern with input signal only. **(b)** Field pattern with input signal plus bias.

lower frequencies, to control the frequency of an oscillator directly and without frequency multiplication. They are much smaller than resonant cavities and have a similar degree of stability. Dielectric resonators are placed in the oscillator feedback path, with coupling being achieved via the magnetic field as shown by Fig. 4.10. Dielectric resonator oscillators (DRO) suffer less frequency pulling than an unstabilised oscillator, a characteristic that can be further improved by the use of an isolator. DROs can be tuned over a range of about 1% (100 MHz in 10 GHz) by perturbing the resonator's magnetic field by varying an air gap between the disc and the circuit's metallic enclosure.

4.7   MICROWAVE CIRCUIT SUBSTRATES AND TRANSMISSION LINES

The printed circuit board concept used for lower frequencies is retained for microwave applications. But the behaviour is now more closely related to transmission lines. Planar construction is used because it is compatible with solid state components and provides low loss and stable circuits, which are very reproduceable and hence cost effective.

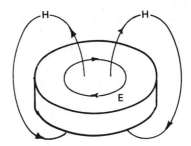

Fig. 4.10 Field patterns in dielectric resonator.

The dielectric substrates used include such proprietary materials as Rexolite, Cuflon and RT Duroid, and PTFE, Alumina and Sapphire. The relative permittivities (ϵ_r) range from around 2.2 to 11.5. The factors that affect the choice of a particular substrate include operating frequency, thermal conductivity, mechanical stiffness, cost, fabrication tolerances and reproduceability, and surface finish which is important for conductor adhesion. For applications above about 5 GHz, alumina is very popular.

The microwave energy propagates in the dielectric substrate and the surrounding air, rather than through the conductors, which simply behave as waveguides. The combined structure then has an *effective relative permittivity* (ϵ_{reff}) somewhere between that for air and the substrate ($1 < \epsilon_{reff} < \epsilon_r$). It is shown in the literature (1) that:

$$\epsilon_{reff} = (\lambda_o/\lambda_g)^2 \tag{4.2}$$

where λ_o = wavelength in free space, and λ_g = wavelength in guiding structure.

Low dielectric constant substrates have the advantage of producing longer circuit elements, which eases the reproduction of dimensional tolerances.

With rising frequency, the propagation tends to concentrate more in the substrate and less in the air, so that λ_g tends to increase with frequency.

Transmission Line Structures

Microstrip

As with most of the structures used at microwave frequencies, the substrate is used to support the metallic conductor patterns. These are commonly gold on a thin film of chromium for better adhesion, and copper. Figure 4.11(a) shows the typical construction of microstrip together with the important dimensions. Figure 4.11(b) indicates how the electric field is set up between the conductor pattern and the ground plane.

The design details for microstrip circuits are quite complex and are best managed using computer aided design (CAD). Design tables are provided in reference (2) and a suitable design program (STRIP) is given in reference (3). In essence, the characteristic impedance of each conductor strip is dependent upon the frequency, the width (w) of the strip, and the height (h) and relative permittivity ϵ_r of the substrate. Each conductor element must

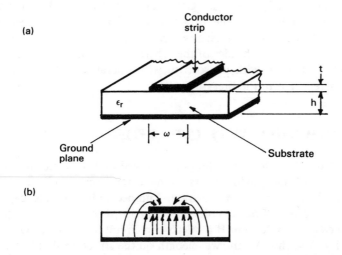

Fig. 4.11 Microstrip waveguide construction and **(a)** important dimensions; **(b)** E field.

be capable of supporting any dc supply current and dissipating any resulting heat. With suitable components and substrate, microstrip can operate satisfactorily up to about 55 GHz.

Inverted Microstrip

In this version, depicted in Fig. 4.12, the substrate simply supports the guiding conductors in air, so that the effective relative permittivity for this structure is practically that of air ($\epsilon_{\text{reff}} = 1$). Thus λ_g is relatively longer than for microstrip, which allows for operation at even higher frequencies.

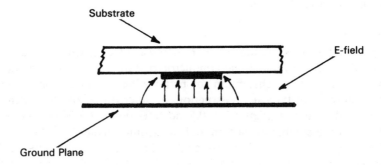

Fig. 4.12 Inverted microstrip.

Fin-Line or E-Plane

This less common structure is illustrated in Fig. 4.13. It consists of a thin dielectric substrate, typically of glass-fibre-reinforced PTFE, metallised with copper and bridging the broad walls of a section of rectangular waveguide. The conductor patterns which form the various circuit elements are defined on one or both sides of the substrate. The losses with fin-line are significantly less than with microstrip, with operation up to beyond 100 GHz being reported in the literature (4).

4.8   SOME MICROWAVE CIRCUIT ELEMENTS

Because of the very short wavelengths involved and the transmission-line-like behaviour of microstrip circuits, many different circuit elements can be fabricated directly on to the substrate. Lumped inductors and capacitors can be provided from open circuit conductor strips. At 10 GHz, λ_o, the free space wavelength, is 30 mm. If a microstrip circuit has an effective relative permittivity of 4, then λ_g, the wavelength on the structure, is 15 mm. Thus $\lambda_g/4$ is only 3.75 mm.

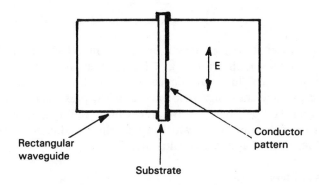

Fig. 4.13 Fin-line or E-plane structure.

Impedance Transformers

A capacitive element can be produced by using an open circuit strip of conductor of length 1, such that $0 < 1 < \lambda_g/4$, whilst an inductive element is produced if $\lambda_g/4 < 1 < \lambda_g/2$. When $1 = \lambda_g/4$, the strip ends both appear resistive, one end of very low value and the other end very high. This impedance transformation allows a quarter-wavelength strip to be used as a matching device. In Fig. 4.14(a), a quarter-wavelength strip of impedance Z_2 is used to provide this match where:

$$Z_2 = \sqrt{(Z_1 Z_3)} \qquad (4.3)$$

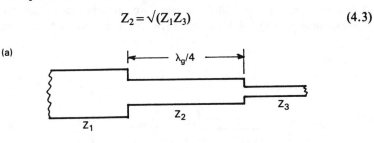

Fig. 4.14 Matching transformer elements.

Z_2 is thus the geometric mean of Z_1 and Z_3. Alternatively, a tapered section of line as shown in Fig. 4.14(b) may be used to affect the impedance match.

Filters

The high impedance strip of length $1 < \lambda_g/4$, connected between two low impedance sections as shown in Fig. 4.15(a), behaves predominantly as an inductor. The two low impedance sections have significant shunt capacitance, so that the combination acts as a low-pass filter (LPF). In Fig. 4.15(b), again $1 < \lambda_g/4$, but this low impedance section is now connected between two high impedances which act as shunt inductors. The overall effect is thus that of a high-pass filter (HPF).

Band Pass Filters

An open circuit stub whose length is a half wavelength is a high impedance at each end. (This is also true for integer multiples of $\lambda/2$.) Thus the half-wavelength stub of Fig. 4.16(a) behaves as a high impedance to signals that make it so. At frequencies on either side the stub impedance falls, so that less signal voltage will be developed. Figure 4.16(b) shows a series of half-wavelength strips (also known as resonators), edge or parallel-coupled via the electromagnetic field. By slightly varying the length of each resonator, the bandwidth of filter can be extended.

Proximity Couplers

A transformer-type coupling between two circuits can be achieved in the manner shown in Fig. 4.17(a). However, such a construction provides only

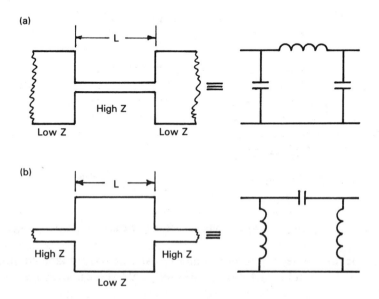

Fig. 4.15 Filters: **(a)** low pass; **(b)** high pass.

(a)

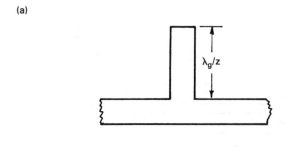

(b)

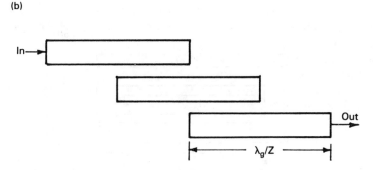

Fig. 4.16 Band pass filters: **(a)** single open circuit stub; **(b)** coupled resonators.

loose coupling. Where a 3 dB coupling between circuits is needed, the 90° hybrid or *Lange* coupler of Fig. 4.17 (b) can be used. This consists of at least four inter-digitated fingers with *air bridge* connections to provide tighter coupling. Due to the length of the coupling structure, power coupled to the signal port is equally divided and transmitted to the coupled and direct ports, but with a phase difference of 90°.

90° Hybrid Branch Line Couplers

When input power is supplied to port 1 of the coupler shown in Fig. 4.18(a), the signal can arrive at port 4 by two paths. One is $\lambda_g/4$ long and the other $3\lambda_g/4$ long. The signals at port 4 are thus anti-phase and self-cancelling, so that port 4 is isolated. The output power divides equally between ports 2 and 3, and since these are separated by a strip $\lambda_g/4$ long there is 90° of phase difference between the signals. The alternative construction shown in Fig. 4.18(b) behaves in the same way because the four ports are spaced by $\lambda_g/4$.

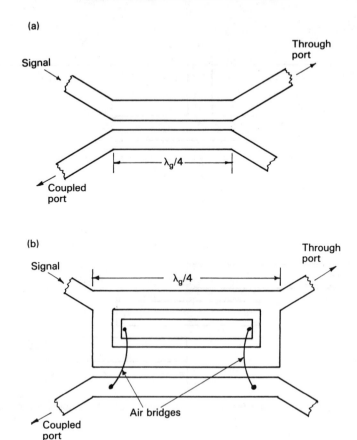

Fig. 4.17 (a) Proximity coupler. **(b)** 90° hybrid or 'Lange' coupler.

Hybrid Ring, or 'rat-race'

This device consists of an annular ring $1.5\lambda_g$ long, with four ports disposed as indicated by Fig. 4.19. Signal power input to port 1 can circulate either clockwise or anticlockwise to reach ports 2, 3 and 4. Both paths to port 4 are $3\lambda_g/4$ long so that port 4 can provide an output. The paths to port 2 are either $\lambda_g/4$ or $5\lambda_g/4$ long, also providing in-phase signals so that port 2 is also an output. At port 3 the path lengths differ by $\lambda_g/2$ to produce anti-phase signals, so that port 3 is isolated. In a similar way, an input at port 3 will not couple with port 1.

Surface Mounted Components

Although not specifically microwave components, the characteristics of these devices is very compatible with the planar technology of microstrip

(a)

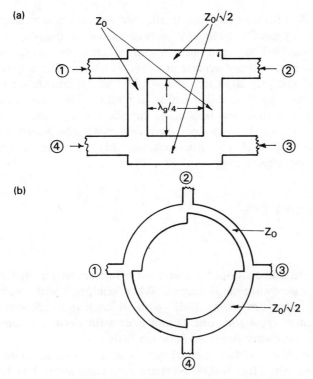

(b)

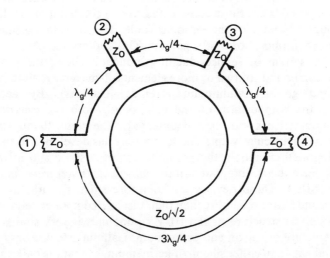

Fig. 4.18 90° hybrid branch line coupler.

Fig. 4.19 Hybrid ring or 'rat-race'.

construction. A full range of these small, leadless devices is available for direct soldering to metallic conductor pads on the circuit boards, which do not need to be drilled for component insertion. These devices are smaller than their conventional counterparts and this leads to smaller sub-systems. Because of the relatively large area of the solder pads, highly reliable joints are produced with good heat-dissipating properties. The components are more efficient and their self-resistance, capacitance and inductance are much lower, so that these devices have better RF characteristics. The technique also lends itself to automated assembly, which in turn further improves the cost effectiveness.

4.9 ACTIVE DEVICES

Amplifiers

Suitably selected silicon bi-polar and field effect transistors (FETs) are available for microwave applications. When equipped with beam lead terminations, they are compatible with microstrip structures. The combination of discrete components and microstrip circuits is sometimes referred to as *microwave integrated circuits* (MIC).

Bi-polar devices are seldom used above about 2 to 3 GHz, whilst metal oxide semiconductor FETs (MOSFETs) are used from about 1 GHz up to around 20 GHz. The ability of MOSFETs to provide voltage and current gain, with relatively lower noise figure and higher efficiency, is a significant advantage.

The most important of the devices that are used as microwave amplifiers up to about 40 GHz are those constructed from a combination of elements from Groups III and V of the Periodic Table. The important elements are aluminium (Al), phosphorus (P), gallium (Ga) and arsenic (As).

The III-V system of semiconductor production allows the growth of cyrstalline compounds of two or more elements. Typically, gallium arsenide (GaAs) and gallium aluminium arsenide (GaAlAs). By accurately controlling the composition of the compound, both the electrical and optical characteristics can be manipulated. The term 'band structure engineering' is particularly apt. The technology has developed to the point where integrated circuits in the normally accepted sense and microwave structures have been integrated into *monolithic microwave integrated circuits* (MMIC). These devices are therefore smaller, with low power dissipation and better reliability than their discrete component equivalents.

The crystalline structure of the III-V compounds is very similar to the familiar structure of silicon and germanium. Gallium arsenide has a higher energy band gap than either silicon or germanium, so that the bulk substrate is capable of being a better insulator. In addition, electron mobility is higher. Together, these features permit a higher frequency and speed of operation, with lower noise levels and improved linearity.

Developments in semiconductor material technology allow the production of very thin *heterojunctions* of GaAlAs and GaAs. The epitaxially grown layer is only a few atoms thick, so that the interface is almost a two-dimensional structure. The electrons which collect in this interface due to the electric field behave differently, because they only have two degrees of freedom of movement. This results in an electron mobility that is a factor of 3 better than that for silicon thus giving rise to devices known as *high electron mobility transistors* (HEMT). Electron mobility is related to the applied electric field by:

$$v = \mu\epsilon \qquad (4.4)$$

where v = average drift velocity cm/sec, μ = electron mobility cm^2/V.sec and ϵ = applied electric field V/cm. HEMT devices can operate at frequencies above 25 GHz, and several types are available with gain and noise factor parameters at 12 GHz that are better than 12 dB and 0.8 dB respectively.

Diode Devices

Diode applications in microwave communications generally fall into two categories: mixers and single-port oscillators. Mixer stages are notoriously noisy and so it is most important to use non-linear devices that are selected for a low noise factor. For some time, the Schottky Barrier diode, with its low threshold of 0.3 volts, has been popular. However, the more recently developed Mott diode is less noisy and is designed as a surface mounting chip. Without beam lead connections, these have less self-capacitance and inductance, which allows them to operate efficiently at least up to 40 GHz.

Common devices in the single-port oscillator group are the Gunn device and the IMPATT (Impact Avalanche and Transit Time) diode. Both devices have a significant negative resistance section in their V/I characteristics. When this negative resistance is placed in parallel with a small positive resistance, say that of a resonant circuit or cavity, oscillations can occur. The origin of the negative resistance is the transfer of electrons from the main conductor energy band into a satellite band, depending on the magnitude of an applied electric field. This results in a pulse of current flowing through the device from cathode to anode.

Table 4.3 shows the typical peak range of values for both devices.

Hot Carrier Diodes

Although these devices are not specifically microwave types, many find applications in microwave systems. A *hot carrier* is an electron that has an

Table 4.3

	Power output	Maximum frequency	Operating voltage
IMPATT diode	1 W	20 GHz	120 V
Gunn device	300 mW	45 GHz	12 V

energy level higher than that of the crystal structure, i.e. several kT above the Fermi level. A rectifying junction can be formed between a metal and a semiconductor (Schottky diode). If forward biased, majority carriers (electrons in N type semiconductor) with an energy level greater than the Schottky barrier – the hot carriers – can cross the barrier and produce a current flow. The number of hot carriers increases with the forward bias so that the forward current rises rapidly. This feature considerably increases the forward sensitivity of the diode compared to that of the conventional PN junctions. The reverse biased current is negligibly small and the diode can switch state very rapidly. These properties make the device valuable for handling very low-amplitude, very high-frequency signals.

Gunn Effect Mechanism

When the applied voltage across the device creates an electric field in excess of about 35 V/cm, a high-level electric field domain is produced in the cathode region, which rapidly drifts to the anode. As the domain passes out at the anode the electric field falls rapidly, only to start to rise again almost immediately. A new domain forms when the field rises above the critical value, and the cycle repeats.

The equivalent circuit of the Gunn device is a negative resistance of about -5 ohms in series with a capacitance of about 0.2 pF. To make the circuit oscillate, the load connected to the device should have:

(1) A series reactance that tunes the device to resonance.
(2) A series load resistance less than the modulus of the negative resistance.
(3) At frequencies other than resonance, the load resistance should be greater than the modulus of the negative resistance.

Provided that points (2) and (3) can be met, a variable reactance can be used to tune the circuit over an octave frequency range.

REFERENCES

(1) Edwards, T.C. (1981) *Foundations of Microstrip Circuit Design*. London: John Wiley and Son.

(2) (1971) *Microwave Engineers' Handbook. Vol. 1.* Artech House 1971.
(3) Slater, J.N. and Trinogga, L.A. (1985) *Satellite Broadcasting Systems: Planning and Design.* Chichester: Ellis Horwood.
(4) (1982) *E-Plane Millimetre-wave Components and Sub-assemblies. M82-0097.* Mullard UK Ltd.

Chapter 5
Digital Signal Processing

5.1 ANALOGUE AND DIGITAL SIGNALLING COMPARED

Wideband communications applications such as television traditionally use analogue signal processing, primarily for reasons of bandwidth conservation and the fact that such systems are well understood. However, each application tends to become unique in certain ways. When the systems become concentrated into integrated circuits, these devices are dedicated to a particular application; relatively few are produced and so their cost is higher than would be the case for mass-produced devices. When the analogue signals are converted into digital form for processing and then back again for such purposes as display, the only dedicated chips are those associated with the interface between the two types of signal. The digital signal processing (DSP) area of the system then uses standard digital ICs that are very much more cost effective.

The increased transmission bandwidth for the digital signal is available on satellite communications links and this overhead is in any case offset by the considerable advantages gained by changing to the digital concept. The systems become more flexible, systems integration can be achieved and computer control gives rise to the concept of *integrated services digital network* (ISDN), where many services can be accommodated with equal performance.

The principal benefits of digital signal processing can be summarised as follows:

(1) More appropriate for linking devices that operate in the digital mode.
(2) Provides transmission speeds that are significantly higher than those commonly achieved with analogue processing.
(3) Provides for improved transmission quality in noisy environments. The noise effects can be reduced using signal regenerators and by using error detection/correction techniques.
(4) More compatible with digital switching techniques used for distribution, and is a natural technique to apply on systems that are linked by optical fibres.
(5) Encryption/decryption can easily be adopted for security of information.

(6) Signal compression techniques (bit rate reduction) can be used to minimise the transmission bandwidth requirements.

(7) For many applications, time division multiple access (TDMA) can be used over the satellite link and this is more efficient in the use of the service than frequency division multiple access (FDMA).

(8) The use of onboard microprocessor-controlled satellite switching becomes feasible and the signal can be regenerated before retransmission to improve the received signal quality. If this improvement is not needed, then regeneration allows smaller antennas to be used.

5.2 SAMPLING AND QUANTISATION

There are several ways of converting an analogue signal into digital form and all are effectively based on a sampling technique. It is shown by the Nyquist theorem that provided a complex analogue signal is sampled at a rate at least twice that of its highest frequency component, then the original signal can be reconstructed from these samples without error.

Figure 5.1(a) depicts the general principle for generating what is commonly known as pulse code modulation (PCM). The analogue signal is sampled at very precise intervals of time, to measure its amplitudes. Since only discrete levels are allowed in a digital signal, each of the amplitudes is quantised or allocated a value which is the integer part of the sampled value (i.e. the integer value of the level that the amplitude is just greater than). These values are then converted into a corresponding binary sequence for digital processing. For the waveform shown in Fig. 5.1(a), the quantised and binary codes series would be:

$$4, \quad 5, \quad 5, \quad 5, \quad 5, \quad 4, \quad 3, \quad 2, \quad 2, \quad 3, \quad 4 \ \ ...\text{etc.}$$
$$100, \ 101, \ 101, \ 101, \ 101, \ 100, \ 011, \ 010, \ 010, \ 011, \ 100 \ ...\text{etc.}$$

It will be noted that each of the eight levels shown can be coded using three bits ($2^3 = 8$).

Any analogue signal that is reconstructed from such a series will obviously be an approximation of the original, the difference or error being referred to as *quantisation noise*. It is shown in Appendix A4.1 that the signal-to-quantisation-noise (SQNR) for a dc signal such as video luminance is given by:

$$SQNR = (10.8 + 20 \log M) \text{ dB.} \tag{5.1}$$

or:

$$= (10.8 + 6n) \text{ dB} \tag{5.2}$$

where M = number of sampling levels, and n = number of bits per code sample. For the example shown in Fig. 5.1(a);
SQNR $= (10.8 + 20 \log 8)$ dB $= 29$ dB.

(a)

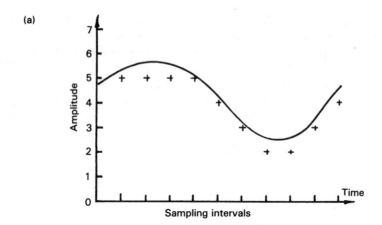

(b)

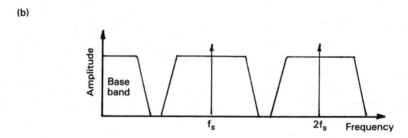

Fig. 5.1 (a) Sampling an analogue signal. **(b)** Spectrum of sampled signal.

Equations 5.1 and 5.2 show that quantisation noise can be reduced to any low level desired by increasing the number of sampling levels and hence the length of the binary code series. The overhead paid for this is the bandwidth of the sampled signal, which can be calculated as follows:

$$2nf_m \qquad (5.3)$$

where f_m is the maximum frequency component in the signal, and n is the number of bits per sample.

The process of sampling produces a frequency spectrum similar to that of amplitude modulation but with an infinite range of harmonics, as depicted by Fig. 5.1(b). The reconstruction or demodulation circuit must contain a low-pass filter to separate the baseband component from the harmonics. If the sampling frequency is not high enough, or the filter cut-off not sharp enough, interference from the first lower sideband will result. Figure 5.2 shows the effect of using a sampling frequency less than $2f_m$, where it is impossible to distinguish between some of the components of the baseband and the lower sideband. Such interference is known as *aliasing*.

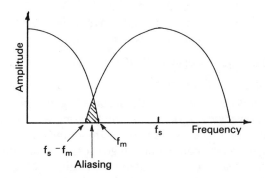

Fig. 5.2 Introduction of aliasing.

5.3 NON-LINEAR QUANTISATION

Reference to Fig. 5.1(a) will show that linear quantisation produces proportionately more noise with smaller amplitude signals. In addition, large amplitude signals are better able to mask the effects of noise. This imbalance can be improved by using a non-linear form of quantisation that behaves in the same way as companding. Such a characteristic, which is relatively easy to implement, is shown in Fig. 5.3 and in the European Standard is referred to as an A-law compander, 'A' being a constant chosen to suite the amplitude distribution on the signal to be processed and typically equal to 87.6.

The curve has a linear region to handle low-level signals and a logarithmic region for the larger amplitudes. Using normalised values, the law for this curve is specified by:

$$y = Ax/(1 + \ln A) \text{ for } 0 \leq x \leq 1/A \tag{5.4}$$

and:

$$y = (1 + \ln Ax)/1 + \ln A) \text{ for } 1/A \leq x \leq 1 \tag{5.5}$$

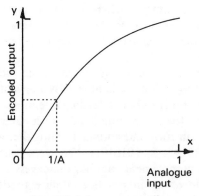

Fig. 5.3 A-law companding.

An alternative compander that finds favour in the North American standard is the μ law device. This is defined by:

$$y = (\log(1 + \mu x))/(\log(1 + \mu)). \quad (0 < x < 1) \tag{5.6}$$

Where μ is typically equal to 255.

It is neither truly linear nor logarithmic, but is a good approximation at the extremes. Whichever device is used at the transmitter, the complementary device must be used at the receiver to ensure an overall linear response.

In modern systems the compander stage is often incorporated within the same integrated circuit (IC) as the analogue to digital (A/D) and digital to analogue (D/A) convertors. These devices are then often referred to as application specific ICs (ASICs). Since the companding is now carried out in the digital domain, the two laws become modified as:

$$\text{A Law: } y = 0.18(1 + \ln(A|x|))\text{sgn}(x), \text{ for } 1/A < |x| < 1$$
$$\text{and} \quad y = 0.18(A|x|)\text{sgn}(x), \text{ for } 0 < |x| < 1/A \tag{5.7}$$
$$\text{u Law: } y = 0.18\ln(1 + \mu|x|)\text{sgn}(x) \tag{5.8}$$

where   x = normalised analogue signal and encoder
                        input or decoder output.
                   y = normalised digital signal and encoder output
                        or decoder input.
                   A = 87.6
                   u = 255.

5.4   QUANTISING AC SIGNALS

The methods of sampling and quantising previously described work well for signals that have a large dc component (such as the luminance signal of television). However, signals with both positive and negative excursions, such as audio, need an alternative approach. One possible way involves using an *offset binary* technique, where a constant is added to each sampled value. But in certain cases, such as with audio mixers, where it is necessary to add signals from different sources, the sum can overflow or exceed allowable peak values.

The commonly adopted solution involves using the two's complement method of representing a binary number. By convention, a leading zero indicates a positive number, whilst a leading one indicates a negative value. It will be recalled that the two's complement of a binary number is formed by inverting each bit in turn and adding 1. Thus the two's complement of 01010101 = 10101010 + 1 = 10101011. When reconstructing the analogue signal from a two's complement sample, the excess 1 should be removed before inversion. However, in practice, failure to do so causes such a minute error that it is often neglected.

Typically, then, an 8 bit two's complement code allows for $\pm 2^7 = \pm 128$ sampling levels.

For ac signals, the RMS signal to quantisation ratio is more applicable. For this case, Appendix A4.1 shows that the SQNR is given by:

$$(1.76 + 6.02n) \text{ dB.} \tag{5.9}$$

5.5 PULSE SPECTRA

Figure 5.4(a) shows a single rectangular pulse of amplitude V and time duration t. Such a pulse can be analysed by transforming it into a mathematical model in the frequency domain. In the time domain, the pulse is defined by:

$f(t) = V$ for $|t| < t/2$ and
$f(t) = 0$ for $|t| > t/2$.

Transformation to the frequency domain by the Fourier transform:

$$F(\omega) = (Vt \sin(\omega t/2))/\omega t/2) \tag{5.10}$$

where $\omega = 2\pi f$ the angular velocity.

A section of the spectrum for equation 5.10 is shown in Fig. 5.4(b), and this is described generally as a sin x/x function. Figure 5.4(c) shows a series of such pulses which is typical of the situation at a receiver. As the zero crossings of the tails occur at the bit cell centres, the energy in the tails tends to be self-cancelling. This is particularly true if the receiver clock circuit is accurately timed. Weak synchronism between the receiver clock and the bit stream will lead to decision errors by the detector circuit, a situation which is often made worse by a relatively long string of zeros.

If the pulse stream were filtered before processing, some of the HF energy would have been removed and the tails of the spectrum would have less significance.

5.6 BASEBAND CODE FORMATS

The errors affecting data services via satellite communication links are chiefly characterised by added white noise and an overall propagation delay that is in the order of 250 mS. The noise leads to the concept of a *bit error rate* (BER), whilst delay variation creates timing and synchronism problems, which in turn lead to further bit errors. As satellite transponders tend to be power rather than bandwidth limited, Shannon's channel capacity rule becomes important.

$$C = B \log_2(1 + S/N) \text{ bits/sec} \tag{5.11}$$

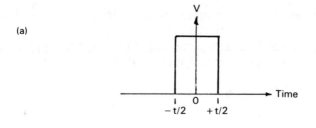

(a)

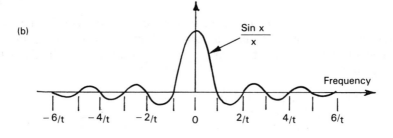

(b)

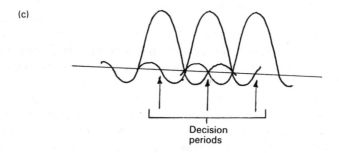

(c)

Fig. 5.4 A rectangular pulse and its spectrum.

where C = channel capacity, B = bandwidth, and S/N = S/N ratio.

The balance between bandwidth and S/N ratio can be used to good effect to maximise the channel capacity for an acceptable bit error rate suitable for the particular service.

Binary code formats are designed to insert extra bits into the data stream on a regular basis, in order to take advantage of this trade-off. A few of the many ways of achieving this are shown in Fig. 5.5, the aim being to minimise the number of consecutive similar bits in the transmitted data stream. The receiver detector clock circuit can then be synchronised to a greater number of signal transitions and so improve its timing. An extra advantage may accrue because some formats have no dc component in their power spectrum, a feature that reduces the low frequency response requirement of the receiver and allows ac coupling circuits to be used.

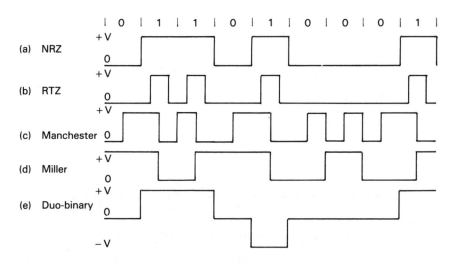

Fig. 5.5 Some baseband code formats.

All the commonly adopted code formats can be generated and decoded using dedicated integrated circuits (ICs), the codes generally being produced from non-return-to-zero (NRZ) basic codes. The return-to-zero (RTZ) format is little used, as its half-width pulses represent an energy/bit penalty.

Codes of the nBmB type are often used to convert n bits/symbol into m bits/symbol (m > n), ensuring that, on average the number of ones and zeros being transmitted are equal.

The bi-phase series of formats for which the basic Manchester code shown in Fig. 5.5(c) is representative has the following features. A signal transition occurs at each bit cell centre, so that a zero is represented by 01 and a one by 10. This ensures that there are never more than two identical bits in series. Another variant of this is the *code mark inversion* (CMI) format, where 0 = 01 and 1 = 00, or 11 alternatively. Although bi-phase codes have a 50% redundancy and double the transmission bandwidth, there is no dc component in the power spectrum as depicted by Fig. 5.6.

The Miller format, shown for comparison purposes in Fig. 5.5(d), is not usually used for transmission purposes. It finds favour in the magnetic storage media. A one is represented by a transition at mid-symbol and a zero by no transition. Except for two consecutive zeros when an extra transition is introduced at the end of the first zero, the Miller code has similar properties to the NRZ code (see section 1.9).

The *duo-binary* code is a bi-polar, full-pulse-width code, as shown by Fig. 5.5(e). It has a bit repetition rate that is half the frequency of the original signal. Zero is represented by 0 volts and one by ±V. The V polarity is unchanged if the one follows an even number of zeros and reversed if it follows an odd number.

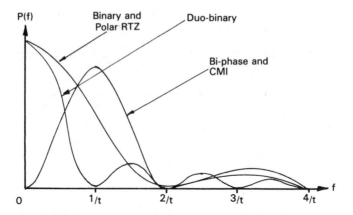

Fig. 5.6 Power spectrum of some code formats.

The more efficient codes use the least redundancy and there is, in general, a trade-off between the complexity of balancing the number of consecutive similar bits and the added redundancy.

The increase in bit rate that these codes produce represents a power penalty. For example, using a 3B4B format involves a bit rate increase by a factor of 4/3 which in turn requires a power increase of about 1.25 dB.

A simple form of error monitoring/detection takes the form of counting the running average of the number of ones and zeros. Any variation outside some predetermined bounds can be used to signal an error situation. With added redundancy, error detection/correction can be provided using *parity checks* or *Hamming codes*.

5.7 DATA ERROR CONTROL (FORWARD ERROR CORRECTION)

When data errors are detected in a bit stream, there are various corrective measures that can be applied. The actual method used often depends upon the source of the original data signal. For instance, had the source been analogue in form, then one of the error concealment techniques might be applicable. This possibility includes:

(1) Ignore the error and treat it as a zero level signal.
(2) Repeat the last known correct value.
(3) Interpolate between two known correct values.

This latter method is really only suitable where a significant amount of data storage is available at the receiver, to allow time to regenerate the analogue signal after processing. In other cases there is the possibility of requesting a repeat transmission when errors have been detected. However, in general this is wasteful of both time and frequency spectrum.

The ASCII code (American Standard Code for Information Interchange) is a commonly used method of representing alphanumeric characters in a digital system. This 7 bit code allows for $2^7 = 128$ different alphabetic, numeric and control characters. The commonly used word length is 8 bits (1 byte), so that space is available for one extra redundant bit.

Even and Odd Parity

A single-error detection code of n binary digits is produced by placing $n-1$ information or message bits in the first $n-1$ positions. The nth position is then filled with a 0 or a 1 (the *parity bit*), so that the entire code word (or code vector) contains an even number of ones. If such a code word is received over a noisy link and found to contain an odd number of ones, then an obvious error has occurred. Alternatively, a system might use *odd parity*, where the nth bit is such that the code word will contain an odd number of ones. In either case, a parity check at the receiver will detect when an odd number of errors has occurred. The effects of all even numbers of errors is self-cancelling, so these will pass undetected. The even or odd parity bits can be generated or checked, using Exclusive OR or Exclusive NOR logic respectively.

Such an arrangement of bits is described as an (n,k) code, n bits long and containing k bits of information. It thus follows that there are $n-k = c$ parity or protection bits in the code word. The set of 2^k possible code words is described as a *block code*. It may also be described as a linear code if the set of code words form a sub-space (part) of the vector-space (entirety) of all possible code words.

Hamming Codes

Error correcting codes have been devised and are named after R.W. Hamming (1), the originator of much of the early work on error control. For a code word of length n bits, it is possible to identify n ways in which a single error can occur. Including the possibility of no error, there must be $n+1$ different code patterns to be recognised.

Hamming showed that the number of parity bits c required for a single error correcting code was given by:

$$c \geq \log_2 (n + 1) \text{ or } 2^c \geq n + 1 \qquad (5.12)$$

(Hamming also defined a 'perfect' code for the case where $2^c = (n+1)$.)

Figure 5.7(a) shows how the message is expanded with redundant check bits. These are usually interleaved with the message bits and placed in positions 2^0, 2^1, 2^3, etc. in the encoded bit pattern. The mechanics of the encoding and decoding process is explained in Table 5.1 for a (7,4) block

(a)

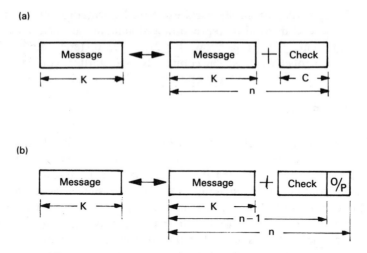

(b)

Fig. 5.7 Hamming code formats (a) SEC; (b) SEC/DED.

code (block length $n = 7$, message length $K = 4$, parity bits $c = 3$). The message bits to be transmitted (0011) are placed in the 3rd, 5th, 6th and 7th positions respectively. The three parity checks are carried out, to determine the values (0 or 1) to be placed in positions 1, 2 and 4. In general, the xth parity check bit is given by the sum modulo 2 of all the information bits, where there is a 1 in the xth binary position. The code word to be transmitted is thus 1000011. If this is now received over a noisy link as 1000001, there is an error in the 6th position. The receiver decoder again carries out the three parity checks and generates the following results:

> 1st check 0
> 2nd check 1
> 3rd check 1

The reverse of the series is called the *syndrome* (syndrome being a medical term for the symptoms of a disease) and this points to the bit in error (110 = 6). Correction is simply achieved by inverting the bit in error. An all correct transmission would have yielded an all-zero syndrome.

By adding an overall parity (O/P) check bit as shown in Fig. 5.7(b), the single-error correction capability is extended to double-error detection, the error patterns being indicated by the following conditions:

(1) No errors – zero syndrome and overall parity satisfied.
(2) Single error (correctable) – non-zero syndrome and overall parity fails.
(3) Double errors (non-correctable) – non-zero syndrome and overall parity is satisfied.

Table 5.1

(Binary)		001	010	011	100	101	110	111
Position		(1)	(2)	(3)	(4)	(5)	(6)	(7)
		P	P	M	P	M	M	M
Check 1		•		•		•	•	•
Check 2			•	•			•	•
Check 3					•	•	•	•
Message				0		0	1	1
Parity bits		1	0		0			
Transmitted Code word		1	0	0	0	0	1	1
Received Code word		1	0	0	0	0	0	1
Recheck of parity		0	1		1	(reverse bit order)		
Syndrome = 110 = 6								

Error in position 6. Invert bit 6 to correct error.

Maximum likelihood Decoding

The *distance* between any two n symbol code words is defined as the number of positions of bit difference. The *minimum distance d* between any valid code words is the minimum number of bit changes necessary to convert one code word into another. For the correction of all combinations of t or fewer errors, the minimum distance must be at least $2t + 1$, thus ensuring that if t errors occur then the received code vector will be nearer to the one transmitted than any other (i.e. $d \geq 2t + 1$).

This concept leads to a simple method of error correction. If a code word is detected as being in error, the decoder checks the distance between the received vector and all valid code words and selects the one exhibiting the minimum number of errors.

Cyclic Codes

In certain transmission systems, the channel noise causes errors to occur not just randomly but in clusters or bursts. A sub-class of the linear block codes valuable in combatting this problem is the group of *cyclic codes*. These have a form such that if a code vector such as 0110 is a valid code word, then so are all its cyclic translates such as 1100, 1001, 0011, etc. which are obtained by shifting the binary sequence one bit at a time to left or right. These codes can easily be encoded or decoded using ICs based on feed-back shift registers.

The *cyclic redundancy checking* (CRC) error detection concept is most effectively described in algebraic form using modulo 2 arithmetic. The basic rules are that $1+1=0$ and $1=-1$, so that addition does not involve a carry and produces the same result as subtraction.

Message and redundant check bits are expressed by the use of polynomials in terms of a dummy variable X, the lowest-order term X^0, representing the least significant bit (LSB) and the highest-order term X^n the most significant bit (MSB). Since the terms range from X^0 to X^n, there will be $n+1$ bits in the message. The coefficients of the terms of the polynomial indicate whether a particular bit is 0 or 1. For example, the 5 bit message stream 11010 would be represented by:

$$1.X^4 + 1.X^3 + 0.X^2 + 1.X^1 + 0.X^0 \text{ or } X^4 + X^3 + X$$

The data stream is written with the MSB on the left when this is transmitted first. The degree of a polynomial is the power of the highest-order term, which in this example is 4.

To generate the code for transmission, three polynomials are used: the message polynomial $k(P)$, a generator polynomial $G(P)$ which is selected from a group of *primitive polynomials* to produce the desired characteristics of block length and error detection/correction capability, and a parity check polynomial $c(P)$. The coded word length $n = k + c$, as for block codes.

During encoding, $k(P)$ is first loaded into a shift register and then multiplied by X^c to move the message c bits to the left, thus making room for c parity bits. The shifted polynomial is then divided by the generator polynomial to produce a remainder that forms the parity check polynomial $c(P)$.

$$(k(P).X^{n-k})/G(P) = Q(P) + c(P) \tag{5.13}$$

where $Q(P)$ is a quotient polynomial and $c(P)$ the remainder which is then loaded into the remaining shift register cells.

The transmitted code vectors for a (7,4) cyclic code $(k=4, c=3)$ using the generator polynomial of $X^3 + X + 1$ are formed as shown by the following example:

Message code $= 1101$, $k(P) = X^3 + X^2 + 1$,
$k(P).X^c = (X^3 + X^2 + 1)X^3$,
$\qquad = X^6 + X^5 + X^3$. Now divide by $G(P)$.

$$
\begin{array}{r}
X^3 + X^2 + X + 1 \\
X^3 + X + 1)\overline{X^6 + X^5 + X^3} \\
\underline{X^6 + X^4 + X^3} \\
X^5 + X^4 \\
\underline{X^5 + X^3 + X^2} \\
X^4 + X^3 + X^2 \\
\underline{X^4 + X^2 + X} \\
X^3 + X \\
\underline{X^3 + X + 1} \\
1 = \text{REMAINDER}
\end{array}
$$

$X^6 + X^5 + X^3 + 1$ therefore divides exactly by G(P), and so forms the transmitted polynomial 1101001. The first four bits 1101 are the original k(P) whilst the remaining three bits 001 are c(P), the parity polynomial. The full list of the $2^k = 16$ code words are shown in Table 5.2.

If a code word T(P) is transmitted, and received without error, T(P) divides exactly by G(P), leaving a zero remainder. The last c bits are then stripped off to leave the original message code. If, however, an error occurs, then division leaves a remainder polynomial that forms the syndrome. There is a one-to-one relationship between this and the error pattern, so that correctable errors can be inverted by logic circuits within the decoder.

The effectiveness of cyclic codes depends largely upon the generator polynomial. For a polynomial of degree n, the decoder is generally capable of detecting/correcting error burst of n or less bits (some errors outside of this bound may be detectable), and odd numbers of random errors. A further advantage in cases where speed is not important is that CRC can operate with microprocessor-based coding, where the system characteristics can be reprogrammed.

Table 5.2

Message vector	Parity vector
0000	000
0001	011
0010	110
0011	101
0100	111
0101	100
0110	001
0111	010
1000	101
1001	110
1010	011
1011	000
1100	010
1101	001
1110	100
1111	111

Golay Codes

These are a sub-set of cyclic codes. The (23,12) versions using 11 parity bits are capable of correcting any combination of three random errors (including a burst of three) in a block of 23 bits. The codes are based on the generator polynomials:

$$G_1(P) = X^{11} + X^{10} + X^6 + X^5 + X^4 + X^2 + 1 \text{ and}$$
$$G_2(P) = X^{11} + X^9 + X^7 + X^6 + X^5 + X + 1$$

both of which are factors of $X^{23} + 1$. Encoding and decoding can be accomplished using ICs based on 11 bit feedback shift registers.

Bose-Chaudhuri-Hocquenghem (BCH) Codes

This very important sub-set of cyclic codes is widely used for the control of random errors. For any positive integers m and t, such that $t \le 2^{m-1}$, codes exist with the following parameters:

Block length $\quad n = 2^m - 1$
Parity bits $\qquad c = n - k \le mt$
minimum distance $d \ge 2t + 1$

which are capable of correcting t or less errors in a block $n = 2^m - 1$ bits. The generator polynomials used must be at least of degree mt. In general, the number of check bits c is almost mt, but if t is less than 5 then $c = mt$.

Values for n, k and t have been tabulated (2) for various polynomials. For example, when $m = 5$, $n = 31$, $k = 26$, $c = 5$, t is equal to 1, but when $m = 5$, $n = 31$, $k = 11$, $c = 20$, t rises to 5.

BCH codes can be processed either by digital logic circuits or by digital computer. The hardware solution is faster but the computer is of course more flexible.

Reed-Solomon Codes

These are a sub-set of BCH codes, which correct any combination of t or fewer errors and require no more than 2t parity bits. The parameters are $n - k = c = 2t$ and $d = 2t + 1$. As with all cyclic codes, a generator polynomial is used for encoding, and the decoder again produces a syndrome that identifies the error pattern.

Two Dimensional Reed-Solomon Codes

These have been devised so that an inner code can be used to correct burst errors of a few bytes length, whilst the outer code can correct very long bursts. In one such code, the data is organised into a matrix of 600 bytes × 30 rows. The first or outer code is then used to add 2 check bytes to each of the columns to create 32 rows. Each row is then divided into 10 blocks of 60 bytes and the second or inner code is used to add 4 check bytes to each block of 60 bytes to give an array of 640 bytes × 32 rows. The inner code is able to correct all single byte errors and tag all errors that are longer than this. The outer code can correct 2 inner code blocks of 60 bytes each, provided that they are tagged when passed to the decoder. It can therefore correct 120 bytes. However, since the inner code is interleaved to a depth of 10 bytes, the outer code, when using the inner code tags, can correct an error up to 1200 bytes long.

Interleaved or Interlaced Codes

This is a simple but powerful way of dealing with either random or burst errors. Any cyclically coded (n,k) set of vectors can be used to produce a new code (αn,αk), by loading coded vectors from the original code into a matrix of n columns and α rows and then transmitting the bits column by column. The transmission sequence is then an interleaved or interlaced code, with an interleaving factor of α. If a burst of errors α or less in length occurs, there will only be one error in each affected word of the original code. As a lower bound, if the original code corrects t or fewer errors, the interleaved code will correct any combination of t bursts each of length α or less.

Because the original code was cyclic, so will be the interleaved one. If the original generator polynomial had been G(X), the generator of the new code would be $G(X^\alpha)$. Therefore the interleaved code can also be encoded and decoded in a similar manner to cyclic codes.

An extension of the interleaving technique is used in certain special cases. This involves the generation of two Reed-Solomon codes from the data, and then cross-interleaving the coded bit patterns before transmission. Provided that the encoder and decoder are synchronised, relatively very long burst errors become correctable as well as single random errors.

Convolution Codes

These codes, used for the control of burst errors, are produced by the complex process of convolution of the message stream with a generator matrix. Unlike cyclic codes, where the parity bits for any given block are contained within that block, the parity bits within a block of convolution code check the message bits in previous blocks as well. The range of blocks over which the check bits are operative is defined as the *constraint length m* for the code (a block code might be described as a convolution code for $m = 1$), the parity bits formed on the current block being dependent upon the message bits in the previous $m - 1$ blocks. The code structure as transmitted is represented by Fig. 5.8(a) and has the following parameters:

Block length $= n$
Message bits per block $= k$
Parity bits per block $= c$
Constraint length $n_0 = m$ blocks $= mn$ bits
Message bits within $n_0 = k_0 = mk$ bits

(a)

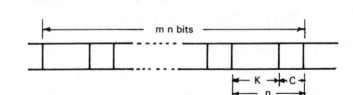

(b)

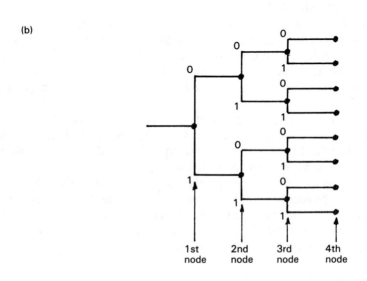

Fig. 5.8 Convolution codes **(a)** Code structure; **(b)** decoding search tree,

Parity bits within $n_0 = c_0 = mc$ bits.

For convolution codes, n and k are normally small integers, often less than 5.

The encoder consists essentially of feedback shift registers and EX OR logic. For decoding, there are two strategies that can be used. Logic or algebraic decoding is restricted to the constraint length. This uses an encoder plus logic circuit to recalculate the parity bits and invert any correctable bits in error. The alternative is probabalistic decoding, where the decoder must be able to store more than m blocks. The sequential decoder then makes its decision on one block, based on more than $m - 1$ previous blocks. The decoding follows a predetermined algorithm through a tree structure such as is shown by Fig. 5.8(b). When the decoder detects an error, it backtracks to the previous node and accepts the alternative branch.

Viterbi Decoding Algorithm (5)

The decoder calculates the distance function between each received code word and all the others in the code book. The word that meets the minimum distance requirement is selected as the most likely one to have been transmitted and is therefore often described as the *survivor*. The decoder stores the survivors and the associated distances to enable a tree search to proceed in the manner of Fig. 5.8(b). Since decoding represents an iteritive process, the decoder can readily be fabricated on an ASIC to handle most of the standard convolution code formats.

Punctured Codes (8)

At higher data rates, Viterbi/Trellis decoding becomes more complex and hence more costly. *Punctured* convolution codes can achieve higher data rates without this penalty. The technique operates by selectively and periodically removing bits from the encoder output effectively to raise the code rate.

In practice, the output from an adaptive convolutional encoder is passed through an analogue gate circuit in blocks of parallel bits. The switching action of the gates is controlled by the *puncture matrix*, which defines the characteristics of the specific error protection for the bits in an information block.

The puncture matrix has to be available at the receiver decoder either as a hardware element or in software. The decoder then treats the punctured code in the same way as erased bits so that these do not affect the decoder decision making.

5.8 PSEUDO-RANDOM BINARY SEQUENCES (PRBS) (PSEUDO NOISE (PN))

For a series of binary digits to be in random order, each symbol must occur by chance and not be dependent upon any previous symbol. Over a long period, the number of occurrences of ones (n_1) and zeros (n_0) should be the same. In a similar way, runs of 2, 3 or more of each symbol should be equiprobable. Such sequences can be generated using shift registers, as indicated by Fig. 5.9, where the logic state of the switches controls the feedback paths, through modulo-2 adders placed between the serial shift register input and output, the state of the switches being set according to a *characteristic polynomial*. Assuming a 4 bit register with the switches set, $S_1 = S_n = 1$, and $S_2 = S_{n-1} = 0$ (1001), then irrespective of the initial states of the shift register cells, the binary pattern shown in Table 5.3 will be produced. The bit pattern b_n is the required sequence (010,110,010, 001,111), generated on a periodic and cyclic basis. Selecting a new characteristic polynomial will create a new sequence.

Table 5.3

	r_1	r_2	r_3	r_4	b_n
State 1	1	1	1	1	0
2	0	1	1	1	1
3	1	0	1	1	0
4	0	1	0	1	1
5	1	0	1	0	1
6	1	1	0	1	0
7	0	1	1	0	0
8	0	0	1	1	1
9	1	0	0	1	0
10	0	1	0	0	0
11	0	0	1	0	0
12	0	0	0	1	1
13	1	0	0	0	1
14	1	1	0	0	1
15	1	1	1	0	1

There are 2^n possible states for the shift register cells, but the all-zero combination is not valid as this would bring the generator to a halt. Therefore the length of a sequence is $2^n - 1$, the period of repetition being independent of the initial conditions.

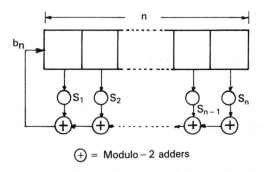

$\oplus$ = Modulo $-$ 2 adders

Fig. 5.9 Pseudo-random binary sequence generator.

It can be shown for these sequences that $n_1 = 2^{n-1}$ and $n_0 = 2^{n-1} - 1$, so that if n is large then the sequence has near-random properties. A maximal length sequence is usually described as an m-sequence and finds many uses in communications systems.

Because of the pseudo-random properties, PRSBs can be used:

(1) As repeatable noise sources for testing digital systems.
(2) To add redundancy to a transmitted data stream, by coding logic 1 as the m-sequence and its inverse as logic 0.
(3) Added to the data stream as shown by Fig. 5.10, to act as a key to ensure secure data.

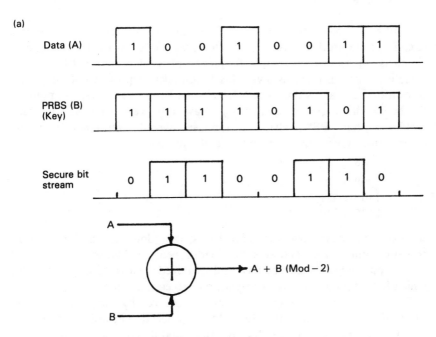

Fig. 5.10 Addition (Modulo-2) of data and m-sequence.

The major difficulties caused by 2 and 3 are synchronism and a reduced data rate.

Gold Code Sequences (6)

These represent a family of pseudo random sequences that were defined by R. Gold in 1967 as having a low cross correlation between each other. As such they can occupy the same signalling channel without producing destructive interference. Because of these properties, Gold codes find applications in code division multiple access (CDMA) or spread spectrum multiple access (SSMA) systems. Each code sequence is produced from a pair of PRBS generators each with n stages by Modulo-2 addition. If all the $2^n + 1$ Gold sequences of period $2^n - 1$ are concatenated, the resulting sequence period becomes $(2^n + 1)(2^n - 1) = 2^{2n} - 1$, the same as would be generated by a shift register with 2n cells.

5.9 BIT RATE REDUCTION/BANDWIDTH CONSERVATION

Pulse Code Modulation

The International Telegraph and Telephone Consultative Committee's (CCITT) standard for pulse code modulation (PCM), as used on telephony channels, operates at a sample frequency of 8 KHz and uses 8 bits per sample (7 bits representing level plus 1 sign bit to represent polarity). The transmitted bit rate is thus 8×8 KHz = 64 Kb/sec. In order to minimise the bandwidth required, quaternary phase shift keying (QPSK) is used. Unlike bi-phase PSK, where each phase inversion represents a bit of information, QPSK uses four phase shifts as follows:

$$0° = 00,$$
$$90° = 01,$$
$$180° = 11,$$
$$270° = 10,$$

so that each phase now represents 2 bits, thus doubling the information content without the increased bandwidth penalty. However, the phase separation between each code symbol is halved relative to bi-phase PSK, which leads to a 3 dB S/N ratio loss, and so companding is used.

Further bandwidth compression can be achieved by using an extension of QPSK. A system known as 16QAM is in operation in which each of the 8 phases separated by 45° can be amplitude modulated to one of two levels. This provides 16 vectors, each one representing 1 or 16 different code patterns each of 4 bits. The bit rate is thus now increased by a factor of 4 without invoking a bandwidth penalty, but at the expense of S/N ratio.

Telephony by PCM is transmitted over satellite links, either by time division multiple access (TDMA) systems, or over digital data channels using the single channel per carrier (SCPC) concept.

PCM can be used for other services such as high quality music or television, but with these wider bandwidth signals the effects of bit errors can be a problem. With PCM, the effect of a single bit in error depends upon its weighting. Whilst an error in the LSB will probably pass unnoticed, an MSB in error is likely to have a significant effect. Using companded PCM on music channels can cause the noise level to vary audibly as the signal level changes.

Delta Modulation

Alternatively, delta modulation (DM) may be used, where the audio signal is simply coded by 1 bit, positive or negative, based upon whether a signal sample is greater or less than the previous value. Because only 1 bit per sample needs to be transmitted, the sampling rate can be increased significantly. This reduces quantisation noise, simplifies the receiver anti-aliasing filter and at the same time reduces the bandwidth required relative to standard PCM. A 1 bit error simply produces a step of the wrong polarity in the output signal, which then retains its correct general shape as shown in Fig. 5.11. Such errors are then likely to pass unnoticed.

An overload effect can occur (also shown in Fig. 5.11) when the input signal amplitude changes by a greater than quantising step size during the sampling intervals, but this chiefly affects large-amplitude, high-frequency signal components, which in audio occur relatively infrequently.

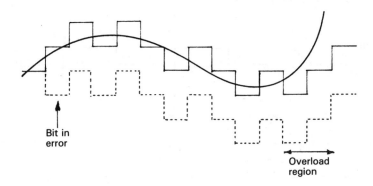

Fig. 5.11 Delta modulation, bit error and overload.

Differential PCM (DPCM)

If the pattern of an analogue signal is known up to some point in time, then because of the correlation between successive values, it is possible to draw certain inferences about future behaviour. This can be achieved by a predictive or extrapolation process.

In predictive coding systems the signal is sampled at regular intervals, but as each sample time approaches the probable value of the next sample is predicted. The difference between the prediction and the actual value is coded as for PCM and transmitted. The receiver must make the same prediction and add the same correction. Figure 5.12(a) shows the basic principles of such a system. However, since the two predictors operate on slightly different signals, this leads to some errors. The system shown in Fig. 5.12(b) is an improvement. Both predictors work on decoded signals and if no noise were added in the transmission channel, both would yield identical results. In the system illustrated in Fig. 5.12(c), the predictor is replaced by an integrator/LPF, so making better use of the previous sample value because of the removal of the ripple due to sampling.

Although the predictions made under any system must contain errors, the advantages claimed for DPCM can be explained as follows.

'In standard PCM, no prediction is made; or in other words, the next predicted value is zero. Since an approximate prediction is better than no

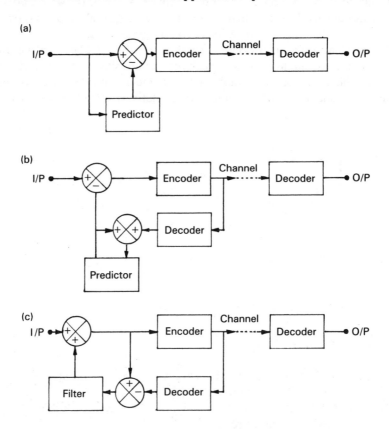

Fig. 5.12 Predictive coding **(a)** basic principle; **(b)** improved version; **(c)** integrator/filter version.

prediction at all and because of the good correlation between successive samples, DPCM must lead to a better S/N ratio as well as a reduced bandwidth.'

Adaptive Systems

Both DM and DPCM systems can give better results if the quantisation step size is made to vary in accordance with the time varying amplitude of the input signal. Such adaptive circuits give a significant improvement in audio quality and behave as companded systems. In addition, pre-emphasis and de-emphasis can be applied to the analogue signals before encoding and after decoding to further improve the S/N ratio.

Sub-Nyquist Sampling

In certain special cases, alternate sampled values are suppressed, so that sampling effectively takes place at half the normal frequency. At the receiver, the missing samples can be replaced by interpolation or predictive coding. The overall effect reduces the bandwidth required by a factor of 2. However, this concept requires the use of a very stable sampling frequency and accurate receiver filtering, plus the extra circuitry for interpolation.

5.10   QUADRATURE MIRROR FILTERS AND SUB-BAND CODING (6 & 7)

It has been shown that to avoid aliasing when digitally processing an analogue signal, the sampling frequency must be at least twice that of the highest frequency component in the complex wave. If aliasing occurs then the original signal cannot normally be extracted (filtered) from the complex spectrum of the sampled signal.

In any wideband signal, many of the frequencies in the band are only rarely present. A good example of this is the television image, where much of the background remains unchanged from frame to frame and with only relatively small areas of movement. By dividing (filtering) the broad spectrum into narrower sub-bands, it will be found that many of these will contain only occasional information. If these sub-bands are sampled, then by using a variable length coding such as Huffman, which allocates the shortest code words to those values that have the highest occurance probability (the stationary areas of the image), a useful bandwidth compression can be achieved.

Quadrature mirror filters (QMF) are basically multi-rate digital filters so that the sampling frequency is not constant throughout the system. Fig. 5.13 shows the basic principle involved with the system containing *decimators* which *down sample* the signal and *interpolators* that provide complimentary *up sampling*.

If a decimator has the digital sequence X(n) as input, it will produce an output Y(n) = X(Mn), where M is an integer that represents the down

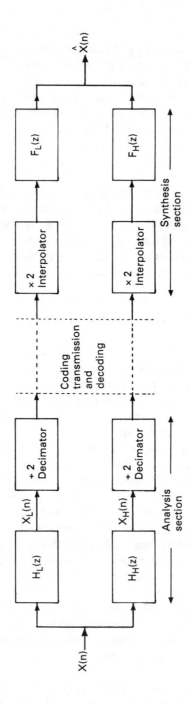

Fig. 5.13 Basic operation of 2 channel QMF system.

sampling ratio. If M = 2, then Y(n) will consist of every second sample. The decimator thus produces a compression in the time domain. In a practical system, before the signal is decimated, it will first be bandwidth limited to reduce the effects of aliasing using a *decimation filter*. At the receiver, the M-fold interpolator will replace the missing M-1 values between adjacent samples. If M = 2, then the interpolator will either average successive values or insert zeros. The effect being a complimentary expansion in the time domain.

The concept is closely related to sub-band coding. As indicated in Fig. 5.13, the input sequence X(n) is divided into separate channels using low and high pass filters with a response of $H_L(z)$ and $H_H(z)$ before down sampling. This signal is then coded for transmission in any suitable manner, before being processed in a complementary manner at the receiver. Higher orders of filter can be produced by further sub-dividing each channel with low and high pass filters. It will be clear from the comments on sub-band coding that some of these channels may contain long strings of zeros, thus allowing significant bit rate reductions to be achieved. The combined response of decimator and interpolator can give rise to aliasing in the frequency domain, but the synthesis filters with response of $F_L(z)$ and $F_H(z)$ remove this from the respective channels. In a correctly designed system, the recombined output sequence $\hat{X}(n)$ will be a perfect reconstruction of the input X(n), even under noisy transmission conditions.

The name quadrature mirror filter, derives from the fact that the response of the filter $H_L(z)$ is a mirror image of that of $H_H(z)$ with respect to the normalised frequency $\pi/2$, which is a quarter of the sampling frequency.

5.11 TRANSFORM CODING (See also Chapter 6, H. 261 Codecs)

This technique, which is commonly used for image signal transmission, is a two step process. First a linear transform is performed on the original image by splitting the space into N × N blocks of *picture elements* or *pixels*. The amplitude coefficient of each pixel is then mapped on to a transform space for coding and transmission. The *discrete cosine transform* (DCT) which is related to the discrete and fast fourier transforms (DFT and FFT) is almost universally adopted. This method is faster because it requires the calculation of fewer coefficients and is more compatible with digital signal processing (DSP) ICs than other transform methods. When used with suitable coding the DCT provides for a high degree of image data compression. This transform technique thus finds applications in videophones, tele/video conferencing, colour fax systems, interactive video discs, and high-definition colour imaging.

Each block of pixel data is transformed using a two dimensional matrix operation in the following manner:

$[T] = [C].[D].[C]^T$ where
$[T]$ is the transformed block,
$[C]$ is the basis of the DCT matrix,
$[D]$ is the original data block and
$[C]^T$ is the transpose of $[C]$.

The DCT coefficients are calculated from the relationship

$$F(u.v) = \frac{4C(u)C(v)}{N^2} \sum_{i=0}^{N-1} \sum_{j=0}^{N-1} f(i,j) \, CosA \, CosB.$$

The inverse transform relationship used at the receiver is given by:

$$f(i,j) = \sum_{u=0}^{N-1} \sum_{v=0}^{N-1} C(u)C(v)F(u,v) \, CosA \, CosB$$

where $A = \dfrac{(2i + 1)\mu\pi}{2N}$ and $B = \dfrac{(2j + 1)v\pi}{2N}$

In both cases:

- i,j and $u,v = 0, 1, 2 \ldots N - 1$ (N being the block size),
- i,j are the spatial coordinates in the original image plane,
- u,v are the corresponding coordinates in the transform plane,
- $C(u) = C(v) = 1/\sqrt{2}$ for $u = v = 0$ and
- $C(u) = C(v) = 1$ for $u = v \# 0$

Such transforms are readily performed within dedicated ICs. The high degree of image data compression available arises not from the transform itself. But due to the fact that many of the transformed coefficients are very small or zero, indicating little block to block variation. In addition, many of the remaining coefficients can be transmitted with lower precision without significantly affecting the received image quality. Since only a relatively few values need to be transmitted, any form of variable or run length coding can be used to advantage.

Depending upon the system requirements, a standard TV image can be reduced to a bit rate as low 64 kbit/s, but 2 Mbit/s is necessary to provide an acceptable quality of colour image.

REFERENCES

Because of the extensive nature of digital concepts, it has only been possible to give a brief description of the more common techniques in use. For a more expansive explanation the reader is referred to works such as those listed in the references below.

(1) Hamming, R.W. (1950) 'Error detecting and Error correcting Codes'. *Bell System Technical Journal* Vol XXVI No.2.
(2) Shu, Lin (1970) *An Introduction to Error-Correcting Codes.* New Jersey: Prentice-Hall Inc.
(3) Halliwell, B.J. (1974) *Advanced Communication Systems.* London: Newnes-Butterworth Ltd.
(4) Cattermole, K.W. (1969) *Principles of Pulse Code Modulation.* London: Iliffe Books Ltd.
(5) Shu, Lin & Costello (1983) *Fundamentals and Applications of Error Control Coding.* New Jersey: Prentice-Hall Inc.
(6) Spilker, J.J. (1977) *Digital Communications by Satellite.* New Jersey: Prentice-Hall Inc.
(7) Vaidyanathan, P.P. 'Quadrature Mirror Filter Banks, M-Band Extensions and Perfect-Reconstruction Techniques.' *IEEE ASSP Magazine*, pp. 4–20, July 1987.
(8) Hagenauer, J. 'Rate compatible punctured convolutional codes'. *Proc ICC '87*, pp. 1032, June 1987.

Chapter 6

Digital & Digital/Analogue Communications Systems

6.1 SYSTEMS OVERVIEW

Practically any digitally based service that can operate within the terrestrial scheme of communications can function over a satellite space link. The services provided range from those specifically designed for computer-to-computer communications, through to the digital processing of services that are essentially analogue in origin.

The principal disturbance to the space link signals is white noise, the effects of which can be mitigated by the use of a suitable error control technique. For digital transmissions, the most significant problems arise from the long propagation delay of at least 0.25 sec overall, and doppler shift. The latter can be a particular problem for high-speed signals and on mobile links.

A significant number of digital services operate through relatively large earth stations, with antennas of 5 metres diameter or greater. In general, these provide point-to-multipoint communications, with the receiving stations acting as hubs for the distribution networks. Many new data services are developing for point-to-point communications, and these receiving stations can operate with antennas of about 1.2 metres diameter or less. It is the latter type of system that this chapter will concentrate on. The systems described are considered as being representative of the much wider range of digital services available.

Digital services are currently provided in the UHF, L, C and Ku bands. (Ka band will probably come into use during the latter part the 1990s.) Bit rates vary from 1.2 Kb/s up to many tens of Mb/s. Many of the earlier, well-established services, operate on the FDMA *single channel per carrier* (SCPC) basis, a significant application being the telephone voice services using various forms of PCM. The satellite multiservice system (SMS) operated by Eutelsat over the European Space Agency's ECS-2 satellite, for instance, divides a transponder's 72 MHz bandwidth into 3200 22.5 KHz frequency slots, each capable of supporting one 64 Kb/s digital PCM channel. A user can be allocated a single slot, or a contiguous multiple of slots, according to the baseband requirement.

Some more recent and developing digital services use the *time division multiple access* (TDMA) system which when combined with *packet switching* and *digital speech interpolation* (DSI) (1), can more than double the capacity of a channel.

TDMA is the complex technique where a number of ground stations time-share a satellite space link in short bursts, often as short as 1 millisecond but sometimes longer. Time-sharing has to be accurately controlled and synchronised by one ground station of the system.

TDMA has much in common with terrestrial *local area networks* (LANs), where the data is organised into *packets* of bits. The make-up of a typical packet structure is shown in Fig. 6.1, where:

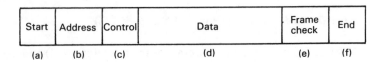

Fig. 6.1 Typical packet structure.

(a) Start flag indicates the beginning of the packet and can contain a *preamble* sequence of bits designed to aid synchronisation.
(b) Each packet carries the originating and destination addresses.
(c) Control section holds the packet sequence number, which is needed if the message occupies more than one packet.
(d) The data bits may include own error protection.
(e) Frame check provides error detection on the packet structure.
(f) An end flag signifies last bit of packet.

Services that handle data in this form are described as being *packet switched*, as opposed to the alternative of being circuit switched, where two communicating terminals can dominate a channel. Many of these services use the CCITT X.25 Packet Radio standard and protocol.

If the TDMA time slots are used on an unallocated basis, some form of *carrier sense multiple access* with *collison detection* (CSMA/CD) must be used. Basically, all the transmitting stations listen continually for a break before starting transmission. If two transmit simultaneously, each will detect the collision of data as noise and relinquish the channel for a random period before attempting to retransmit.

Between the extremes of FDMA and TDMA, there are a number of hybrid systems. These vary between using different transmission modes in each direction, to the use of several low-speed data TDMA carriers on one transponder. With the rapid developments in digital signal processing (DSP), *code division multiple access* (CDMA) (see page 7), provides a further convenient way of improving channel occupancy.

Digital speech interpolation (DSI) is used on TDMA systems and makes use of the time gaps that appear in all voice communications channels. The gaps are then allocated to other users waiting to transmit data.

Such schemes allow satellite communications to become integrated into a global system, thus giving rise to the *integrated services digital network* (ISDN) concept.

Whichever way the space link is managed, the general ground receiver will be based on the double or triple superhet principle. The head-end electronics at the antenna act as a low-noise convertor for a block of channels and provide the first intermediate frequency (IF), which is fed over land lines to the main receiver situated some distance away.

Comparative Down-Link Budgets (TDMA and SCPC)

This comparison is based upon data obtained from references (2) and (3) for the EUTELSAT SMS service. The down-link service operates at 12.5 GHz in two possible extreme modes: TDMA at a transmission rate of 24.576 Mb/s (74 dBHz), or FDMA (SCPC) at 64 Kb/s (48 dBHz). Both use QPSK to achieve a maximum bit error rate (b.e.r.) of 10^{-4}. Table 6.1 assumes that the EIRP over the -3 dB service area is at least 42 dBW. But due to power limitations, the transponder can only support 100 (-20 dB penalty) 64 K/bs carriers simultaneously. The receiving ground station is assumed to be equipped with a 2 metre diameter antenna with an overall efficiency of 65%, yielding a gain of 47 dB.

Table 6.1   Down-link power budget

		TDMA	SCPC
(a)	Transponder saturated EIRP (-3 dB contour)	42 dBW	42 dBW
(b)	Back-off for multi-carriers	0	6 dB
(c)	Operating EIRP (a–b)	42 dBW	36 dBW
(d)	Transponder power per carrier	42 dBW	16 dBW
(e)	Free space attenuation	208 dB	208 dB
(f)	Receiver antenna gain	47 dB	47 dB
(g)	Received signal power (d–e+f)	-119 dBW	-145 dBW
(h)	Receiver noise temperature	25 dBK	25 dBK
(i)	Down-link C/T ratio (g–h)	-144 dBW/K	-170 dBW/K
(j)	Boltzman's constant	-229 dBW/HzK	-229 dBW/HzK
(k)	C/N_0 ratio (i–j)	85 dBHz	59 dBHz
(l)	Transmission rate	74 dBHz	48 dBHz
(m)	E_b/N_0 (k–l)	11 dBHz	11 dBHz
(n)	Degradation margin	4 dB	4 dB
(o)	E_b/N_0 (theoretical (m–n)	7 dB	7 dB

The references (2,3) show that there is no significant difference in signal degradation due to the up-link at 14 GHz.

The Down-link analysis of Table 6.1 shows that there is very little difference in terms of E_b/N_o (equivalent to S/N ratio (see page 15)) between the two modes. FDMA is the least efficient in terms of throughput of data because of the power limitation, whilst TDMA makes full use of the available bandwidth at all times. However, TDMA systems are more complex and thus more expensive to implement and maintain.

Table 6.2 Standard FAX groups

		Group 1	Group 2	Group 3	Group 4
Generic title		Low speed	Medium speed	High speed	Ultra high speed
Transmission speed (A4 document)		6 min	3 min	1 min	2–4 sec
Modulation		FM	AM/PM-VSB	DPSK	Digital (64 Kb/sec)
Carrier frequency		1700 Hz ± 400 Hz	2100 Hz	1800 Hz	
White level		1500 Hz	Max Amp		
Resolution	Vertical	3.85 1/mm	3.85 1/mm 5.3 1/mm	3.85 1/mm 7.7 1/mm	7.7 1/mm 15 1/mm
	Horizontal		5.3 pel/mm	8 pel/mm	16 pel/mm
Image signals		Analogue		Digital	Digital
Handshake signals		Audio tones		300 b/sec FSK (CCITT V21)	ISDN compatible (CCITT V29/33)
Redundancy reduction		None	None recommended	READ	Modified READ

6.2 FACSIMILE SYSTEM

This service, originally developed to transmit photographic and documentary information over standard telephone cables, is now used over radio and satellite links. The many varied services available range from the transmission of weather maps and geological mappings from offshore oil platforms to the transmission of complete pages of a newspaper for remote printing. During the last 40 years or so, a number of international standards have been formulated under the auspices of the International Telegraph and Telephone Consultative Committee (CCITT) (4). Table 6.2 lists the main characteristics and differences of the four groups of machine in service. Group 1 terminals are now almost obsolete, with the major part of the service being carried by machines of Groups 2 and 3. Usually these are so equipped that they can communicate with each other. Ultra high-speed Group 4 terminals, which are capable of reproducing an A4-sized document in about 5 seconds, are becoming available to work over digital networks (ISDN). Whilst Group 4 machines are currently (1991) in operation, this CCITT standard still provides significant flexibility to allow for future developments (28).

The information in the document to be transmitted usually lies in dark markings on a light background. This can be analysed by segmenting the document into elemental areas small enough to resolve the finest detail needed. The document for transmission is scanned sequentially by a light beam, in a series of very narrow strips. The magnitude of the reflected light from each *picture element* (pel or pixel) is then used to generate an electrical signal. In earlier machines, this was accomplished by a photocell/electron multiplier tube, but this has now been superseded by an array of semiconductor devices, either photodiodes or charge-coupled image sensors. As Fax service terminals are transceivers, this signal can now be used to construct an accurate facsimile of the original document, either locally or at a distance.

Table 6.2 also shows the several methods of sub-carrier modulation in use. These are necessary in order to accommodate the signal within the 300 to 3400 Hz bandwidth of the telephone channel and to allow ac coupling to be used. In spite of these differences, Group 2 and 3 machines can work together. Whilst the G4 terminals are microprocessor controlled, they should be capable of being programmed to work with G3 machines. Unlike the G3 terminals, the G4 units were originally designed to operate over the wider bandwidth, leased lines service. However a high-speed modem is available so that G4 terminals can operate over the voice channels of the *public switched telephone network* (PSTN). The increase in transmission rate achieved by the G4 terminals is due to an improved data compression algorithm and the availability of the higher data rates on CCITT V29 and V33 systems. These provide for basic data rates of 9.6 or 14.4 kbit/s respectively.

Because the G4 machines are ISDN compatible, this design has been extended to provide a PC−Fax service by the addition of a *fax card* (circuit), to the standard bus system of a personal computer. This concept not only allows for computer to computer communications, but also for services that include E-mail (electronic mail), store and forward or store and retrieval messaging and even optical character recognition (OCR). The system is compatible with the International Organization for Standardisation's (ISO) seven-layer *open systems interconnect* (OSI) standard.

To ensure that the receiving terminal reproduces or *writes* the same corresponding Pel that the transmitter has produced, a *phasing signal* or *white burst* is transmitted, always at the beginning of each page and often at the beginning of each line. It is also important that the writing and reading rates are synchronised. To this end, all timing signals are derived by division from a crystal-controlled oscillator.

It is not always necessary for the facsimile to be the same size as the original. Certain terminals have the facility to enlarge or reduce. However, if distortion in the document is to be avoided, the page aspect ratio should remain constant. The compatibility of aspect ratios is reflected in the factor or index of cooperation (FOC or IOC), which is based on the ratio of scan line length to vertical scanning density (width).

$$\text{FOC} = \text{Effective scan length} \times \text{vertical scan density} \qquad (6.1)$$
{Effective scan length = actual length + 5% to accommodate the phasing signal}

The IOC ratio was originally defined for drum-scanned systems, hence,

$$\text{IOC} = \text{FOC}/\pi \qquad (6.2)$$

Provided that the machines working together have the same ratio, the document shape will be retained even though its actual size may be different.

The standards for resolution, which define the finest detail that can be reproduced, are also shown in Table 6.2, from which it can be seen that the resolution of Group 2 terminals is about 20 pels/mm² (3.85×5.3). As the pel size is reduced, the vertical scan density increases to produce a higher value of IOC (FOC). A Group 3 machine at minimum resolution produces slightly more than 30 pels/mm², so that an A4 page of 210 mm × 297 mm would be resolved into about 1.9 million pixels. If all these have to be transmitted, the coding system used must be very efficient to minimise the transmission time and/or the bandwidth required.

The study of a typical document will show that it contains considerable redundant information which need not be transmitted. Redundancy reduction can work in two dimensions:

(a) Many sections of each scan line are continuously white and so hold no information to print. Omitting these produces a horizontal economy known as one-dimensional coding.

(b) On average, one scan line bears a close resemblance to each of its neighbours. This fact can be exploited by only transmitting information on how the current line differs from the previous one, thus producing a vertical economy.

The use of both horizontal and vertical coding (two-dimensional) is often referred to as *relative element address designate* (READ). Its use can reduce the transmission time by a factor of about 10.

Originally, frequency modulation of the sub-carrier was used because of the noise reduction properties. However, the need to increase speed, and hence bandwidth, caused the adoption of a modified form of AM for Group 2 terminals.

A normal double sideband AM signal can be filtered to remove a part of one sideband, leaving a *vestige* of it to provide a vestigial sideband (VSB) signal. This reduces the AM bandwidth and still leaves a signal that can be processed in a normal AM system. The use of VSB saves about 30% on bandwidth to support a corresponding increase in transmission speed.

A further saving of bandwidth can be effected by the introduction of a phase modulation component. Assume that white represents a peak positive voltage and black zero volts. A series of transitions between black and white thus represent a particular frequency. If alternate white peaks are now inverted to produce a negative voltage (see alternate mark inversion–AMI), the fundamental frequency will be halved. When this signal is used to modulate the sub-carrier, the negative peaks produce a carrier phase reversal, producing a *white* signal without any change of frequency or amplitude. This form of modulation, known as AM/PM-VSB, has a reduced bandwidth such that the transmission speed can be doubled.

An alternative time-saving method causes the system to scan more rapidly when no black/white transitions are detected, i.e. in a white area. This *skipping white spaces* requires a feedback network in the scanner mechanism and is used successfully in the *Pagefax* system that is used to transmit whole pages of newsprint. In transmission, Pagefax occupies twelve telephony channels (48 KHz). The document service, *Docfax*, as transmitted over the Inmarsat satellite links, occupies a normal telephony channel, and is transmitted FM/SCPC for both analogue and digital terminals.

The Facsimile Terminal

Figure 6.2 shows the basic organisation of a facsimile terminal. The document to be transmitted is illuminated in a series of very narrow strips, each of which is scanned from side to side sequentially. The light reflected from the document surface is focused on to a phototransducer to generate the electrical signal. This is then modulated on to the sub-carrier and filtered to ensure that the signal lies within the spectrum of a telephone channel. A line amplifier is used to achieve the correct signal level and impedance match for the transmission medium.

The recording (*writing*) path is practically complimentary. The received signal is first amplified and then filtered to remove any line noise. Demodulation restores the baseband signal, which then passes through a *marking amplifier* to produce the facsimile document.

Figure 6.3 gives an indication of how the facilities can be extended using microprocessor control. Semiconductor technology provides the basis of better resolution and faster transmission. An array of 5000 photodiodes can

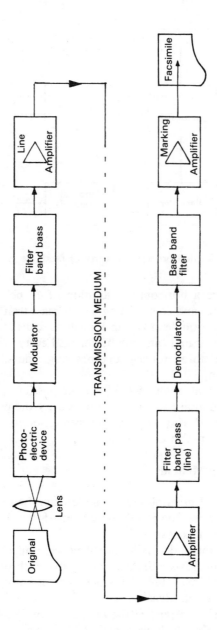

Fig. 6.2 Basic facsimile system.

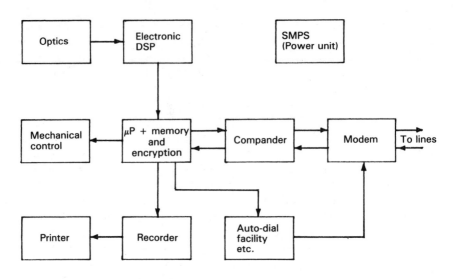

Fig. 6.3 Microprocessor control of facsimile system.

be used to produce a horizontal resolution of 16 pels/mm. The use of semiconductor stores and companding can increase transmission speeds up to 9.6 Kb/s. As an example, a Group 3 machine scans each pixel in about 2 μS, whereas the line signal can only be changed every 140 μS. The scanned signal is therefore stored for signal processing, coding and companding before transmission.

Microprocessor control also introduces the possibility of encrypting the data stream in the digital Fax terminal for security and privacy. The proven technology of *auto dialling and auto answering*, often found in local area networks (LANs) can also be applied.

Recording (writing) Methods

Of the many ways of creating the facsimile from the received signal, only the electrolytic, electrostatic and thermal methods have been popular in the most generally used Fax systems.

The electrolytic method uses a special paper, impregnated to render it conductive. Current flows from a scanning stylus, through the paper to a fixed back contact. This causes the paper to discolour in proportion to the magnitude of the current density.

Variants of the electrostatic method are common in office photocopiers, but less popular in Fax machines. Styli attached to a flexible belt are driven across standard paper to leave a charged pattern on it. The styli voltage varies between + 600 volts for black and − 450 volts for white. The charged paper passes over a magnetic roller bristling with magnetised iron particles and holding a black toner powder in the interstices of the bristles. The powder is attracted to the positively charged areas and repelled by the negative regions. The powder thus forms a facsimile of the original document. Passing the paper over a heated roller causes the toner to bond permanently to the paper.

The thermal method which is now commonly used employs a special thermosensitive paper, which discolours when its temperature exceeds about 70°C. Full black is produced at about 110°C. The paper is passed over an array of minute heating resistors which are made using thin-film technology. This provides a heating element with very little thermal inertia. The elements can thus respond rapidly to the changing signal. Up to 2000 elements may be mounted in an array, giving a horizontal resolution of 8 pels/mm, a figure which for Group 4 machines is likely to double to 16 pels/mm. One end of each element is connected to a critical voltage, whilst the signal voltage is supplied sequentially to the other end. The paper temperature is made to vary between 60°C and 110°C, to produce a copy with varying tonal shades if required in the original.

Copying on to standard paper is also carried out using thermal transfer technology. A special meltable ink film is carried on a base film and this can be transferred to the copy paper by the heated thermal heads. Colour printing is also possible using this technique, by transferring coloured inks (yellow, cyan or magenta), successively overlaid, to provide a range of colours.

Two further methods may be found in Weather Fax recording systems. A photographic unit which uses photosensitive paper, stored either in cassettes or as separate sheets, for automatic feeding, provides the facsimile. The exposure light may be a laser or a high-brightness lamp source. The unit contains facilities to complete the photographic development, and this normally follows on automatically after exposure. The method has the advantage of a high grade of reproduction with a grey scale of up to sixteen levels. However, as the process is essentially a wet one, the equipment is not very suitable for portable operation and the cost of the unit and the paper is high.

A dry silver paper process that can produce results almost as good and with a grey scale of about eight levels is also available. The recording process uses a laser to make the exposure. The paper cost is the highest of all methods and can provide some storage problems. If stored at too high a temperature, the facsimile quality becomes degraded.

Handshaking and Synchronising

Handshaking is the term used to describe the sequence of signalling used between communicating terminals prior to the transmission of fax signals. This procedure is carried out to check the mutal compatibility and status, to issue control commands and, in some cases, to monitor the line conditions. It is a way of identifying each machine's capability, whereupon one will select a particular mode, which in turn will be acknowledged by the other. Following the handshake procedure, the originating station transmits a start signal, the nature of which depends upon the group. A white burst may be transmitted for 30 seconds, or one of two audio tones. This will be followed by a phasing burst and then the document signal proper. Stop may be signified, either by loss of carrier for 5 seconds, or another tone burst.

After the transmission of the document, further handshaking confirms satisfactory reception and indicates whether more pages are to be sent or not.

Applications involving satellite links can present some problems. Handshaking inherently implies a duplex link for communications, and the propagation delay needs to be taken into consideration. In addition, interference and signal fading can introduce a signal corruption.

Weather Chart Recorder

The Muirhead Murfax-949S (5) is a receive-only Fax terminal developed to World Meteorological Organization (WMO) standards for the reception of weather charts from the various satellite sources. The terminal, which is microprocessor managed, is capable of either local or remote control and can process either AM or FM signals automatically. It can thus work with signals from either polar orbiting or geostationary weather satellites. Printout is produced on electrosensitive paper. The unit is made suitable for either fixed, ship-borne or other mobile operations. The organisation of the signal controlling section is displayed in Fig. 6.4.

Image signal input

The satellite image data is fed to the signal programmable interface adapter (PIA). This circuit also contains a multiplexer which switches the buffer, either to the data source or to a test reference, the switching action depending upon the status of the front panel control (local or remote). The digital control bus which is managed by the microprocessor (MPU) program directs these inputs via the control input/output (I/O) PIA to the multiplexer.

Carrier Detection

The buffered signal is applied to the carrier detect and dropout delay circuit. This also contains a tone detector which supplies a *flag* indicating when a carrier is present. The MPU is thus signalled via the signal PIA to start whenever a polar orbiter carrier is present and stop when the carrier

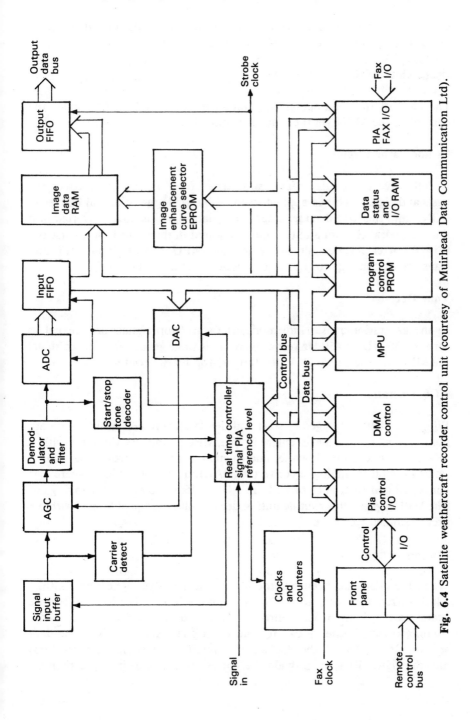

Fig. 6.4 Satellite weathercraft recorder control unit (courtesy of Muirhead Data Communication Ltd).

drops out for more than 3 seconds. The dropout delay circuit is included to avoid a noisy carrier prematurely stopping the system.

Image Data AGC

The level of the data signal is maintained by an AGC loop, under the control of the MPU via the real time controller.

Demodulator Circuit

A full wave demodulator and sub-carrier filter is used to recover the baseband signal. This is applied to the 8-bit analogue-to-digital converter (ADC) and the tone decoder stages. The data signal is now converted into the 8-bit parallel from and temporarily stored in the input first-in-first-out (FIFO) buffer. At the appropriate time, the MPU reads the input FIFO and transfers its contents to the various nodes over the data bus.

Tone Decoder

In the geostationary satellite mode, the start/stop signals are carried as tones of 300 Hz or 450 Hz respectively. This stage generates a flag to signal the MPU via the signal PIA, to start or stop its run routine.

AGC Operation

When the MPU is flagged to start and the data is present on the 8-bit data bus. The AGC routine is entered. The MPU compares the data with a reference based on peak white, and calculates a gain control value. This is sent to digital-to-analogue converter (DAC), as an 8-bit word, where it is converted and filtered to provide an AGC voltage, to adjust the loop gain. The MPU repeats this routine until it decides that the signal level is within a predetermined tolerance.

Sync Detection

When the AGC has been set, the MPU enters its sync detection routine, which differs according to the signal format selected. The MPU therefore examines the front panel control settings to decide which sync software routine to run. It then looks for the appropriate sync words in the data stream. Once detected, the clocks and counters are reset via the control bus and the signal PIA, to provide the correct frame and line sync timings.

Clocks and counters

The main Fax clock provides an input at the very precise frequency of 3.949440 MHz. This is divided down to generate the sampling and data strobing references. The division ratios which are dependent upon the image data format selected produce frequencies of either 7.2 KHz or 14.4 KHz, for polar orbiter or geostationary reception respectively. These frequencies are further divided down to produce either 2 Hz or 4 Hz for the line blanking interval. The output data to the recorder is fixed at 4114 words per line, of which a 185 word period represents the blanking period. Thus each printed line requires 3929 data words which are strobed out and counted.

Image enhancement

Although the recorder carries manual controls for the adjustment of *brightness* and *contrast* of the printout, it also has a facility for *image enhancement* which may be manual or automatically controlled by the MPU. An electrically programmable read only memory (EPROM) carries a selection of enhancement curves, and when one is selected a copy is mapped into the image data random access memory (RAM). Under the direct memory access controller (DMAC), the data in the input FIFO is transferred over the data bus to address the image data RAM, where each address corresponds to a point on the enhancement curve. This point, once addressed, is dumped as data into the output FIFO, from where it is strobed as enhanced output data to the fax printer.

Control Input/Output

Control of the MPU unit and the Fax recorder is undertaken via the control I/O PIA and handled by the MPU to give the required response. The MPU can be programmed manually from the front panel, or remotely via an 8-bit parallel bus through the control I/O PIA.

Fax Input/Output

The Fax I/O bus is used to carry handshake and status signals between the recorder section and the control unit.

6.3   SATELLITE WEATHER AND ENVIRONMENTAL SERVICES

These services were set up under the auspices of the World Meteorological Organization (WMO) to provide a permanent World Weather Watch (WWW) and carry out experiments within the Global Atmospheric

Research Programme (GARP). The total system at any one time involves the services of six geostationary satellites and at least four polar orbiters.

Of the geostationary satellites, two are operated by the USA and are known as Geostationary Operational Environment Satellites (GOES); one, the Geostationary Meteorological Satellite (GMS), is operated by Japan; one is operated by the USSR and is referred to as Geostationary Operational Meteorological Satellite (GOMS); one is operated by the Indian Satellite Research Organisation (ISRO) and known as INSAT; whilst that operated by the European Space Agency (ESA), Eumetsat, belongs to the Meteosat series.

These satellites are supported by about 9000 earth stations, 7000 ships and 850 balloons from more than 150 countries. Unlike the other satellites, GOES-H also monitors the earth's magnetic field, and the intensity of the *solar winds* and the radiation belts around the earth. It also acts as a relay link in the 406 MHz band, for the international search and rescue service, COSPAS-SARSAT.

The geostationary satellites are spaced about 70° longitude apart, in order to provide a wide coverage up to about 75° latitude. The actual orbital positions are:

> GOES-F at 135°W (Pacific Ocean Region W) (POR)
> GOES-H at 75°W (Atlantic Ocean Region) (AOR)
> METEOSAT at 0° (Greenwich Meridian)
> GOMS at 70°E
> INSAT at 74°E (Indian Ocean Region) (IOR)
> GMS at 140°E (Pacific Ocean Region E) (POR)

With at least 1 in-orbit satellite.

The view from these satellites takes in a full earth disc that covers about one-quarter of the surface, although near to the horizon the view is too oblique to be of much value.

The polar orbiting satellites provided by the USA are of the TIROS and NIMBUS series (6) and are operated by the National Oceanographic and Atmospheric Administration (NOAA). The satellites provided by the USSR are of the Meteor series. All these have a period of about 100 minutes, at an altitude of between 700 and 1500 km (7). Due to the earth's rotation, each orbit crosses the equator about 25° longitude further west than the previous orbit. The on-board instruments see the earth surface as a series of strips, which are scanned from side to side. A particular location is thus viewed at least twice a day, once during a north-to-south pass and again, about 12 hours later, during a south-to-north pass. Because of the width of each scan, there is an overlap of tracks, so that some areas are viewed on successive orbits.

The two systems together provide a complete world-wide coverage to detect weather patterns as they develop.

A radiometer that measures the magnitude of earth radiation (heat, light, etc.) is carried on each satellite; sensors convert this data into electrical signals for transmission. To provide the most useful images for meteorological purposes, the radiation waveband is restricted by the use of filters. The Meteosat data that is transmitted (8) covers the following bands or channels:

> 0.4 to 1.1 μm – visible range (VIS).
> 5.7 to 7.1 μm – water absorption range (WV)
> 10.5 to 12.5 μm – infrared range (IR)

to provide three distinct images.

Visual range (VIS): These images show only the reflected light from the earth and so are only available for the daylit hemisphere. In the reproduced image, space appears black and snow and clouds white. Other areas are shown as varying shades of grey. Figure 6.5 is an example of a computer-

Fig. 6.5 Visual image from Meteosat (courtesy of Meteorological Office, Bracknell).

enlarged view of Western Europe (courtesy of the National Meteorological Office, UK).

Water vapour (WV): These images do not provide an earth view and only represent the upper atmosphere. White areas represent low temperatures, which can be equated with high humidity. The darker areas represent a lower level of atmospheric humidity.

Infrared (IR): Infra-red radiation is proportional to temperature, so these images are available during both daylight and darkness. Since space and high clouds are cold, these regions are depicted as white. Hot deserts appear very dark. The grey scale is thus temperature dependent. Figure 6.6 is an example of a processed IR image (courtesy of the National Meteorological Office, UK) – processing has caused the region of space to become black.

Fig. 6.6 Infra-red image from Meteosat (courtesy of Meteorological Office, Bracknell).

Each geostationary meteorological satellite spins on its axis at 100 rmp, and its radiometer scans the earth in the east-to-west direction. After each revolution, the on-board sensors are deflected in small steps in the south-to-north direction. This dual action causes the earth to be scanned in a series of lines, a full image being generated at half-hourly intervals. Each image data is gathered during 1/20 of each revolution, which represents a one-line sampling period of the earth radiation. Each sample contains 2560 32-bit words, each of which contains visual, infrared and possibly water vapour data. The corresponding bit rate can be calculated as follows:

Time for 1 revolution $= 1/100 \times 60$ secs $= 600$ mS
Radiometer sample period $= 600 \times 1/20 = 30$ mS
Total number of bits per sample $= 2560 \times 32 = 81920$ bits
Bit Rate $= 81920/0.030 = 2.730666$ Mb/s

The raw data for each line of the image is stored in memory and then transmitted in somewhat less than 570 mS at 166.7 Kb/s, using digital split phase modulation (SP-L). This is a version of the Manchester code, in which a transition occurs at each bit cell centre. Effectively, $1 = 10$ and $0 = 01$. On Meteosat, the carrier frequency is 1686.833 MHz, linear polarised at an EIRP of 19 dBW per transponder. If the line store fails, provision is made to bypass the 'stretched mode' and transmit the raw data in a 'burst mode' at the same rate that it was gathered (2.73 Mb/s).

To provide a global service, the data is initially processed at centres in Melbourne, Moscow or Washington, the final reports being produced either in Washington (US) or Bracknell (UK).

The data for each image is gathered in 25 minutes and a further 5 minutes is allowed to reset the vertical scanning mechanism, thus providing a new image every 30 minutes. The maximum earth resolution is either 2.5 or 5 km depending upon the image mode. After the raw data has been received and stored at the earth station meteorological centre, it is processed before being retransmitted, through the satellite, in both analogue and digital form for dissemination to the various users. Processing is necessary to remove the defects in the radiometer, sensors and filter characteristics, to correct any inaccuracies in synchronism and to ensure registration of the three views. Processing also allows other derived images to be transmitted, such as cloud motion vectors and cloud top heights.

Apart from earth imaging, the geostationary satellites provide two further services to meteorology: the dissemination of processed raw data to other areas of the world, and data collection via data collection platforms (DCPs). These are small automatic or semi-automatic units that monitor environmental data and use the satellite to transmit this back to a central ground station. These platforms may be carried in ships or aircraft, or be fixed on land sites. Some are self-timed to transmit their data by an internal clock. Some may transmit their data upon interrogation from a central station. An 'alert' DCP automatically transmits a message if one of its measuring parameters exceeds a preset level. This gives early warning of risks of serious disturbances. DCPs are interrogated over one of two channels and transmit within the frequency range 402 to 402.2 MHz.

The DCPs are supported by a system of three geo-stationary satellites designed by NASA to act as data relays. Because these can *see* the polar orbiters for longer periods of time than an earth station, they increase the system coverage with fewer ground installations. The tracking and data relay satellites, (TDRS) as these are known, provide a service to LAND-SAT, the NASA space shuttle, C-band satellites and 19 LEOs. The system operates on several carrier frequencies, using S band for data rates up to 250 kbit/s and Ku band for higher data rates. The down link to the earth station is by Ku band (13.5 to 15.2 GHz) using a spread spectrum technique that provides a multiplex for up to 30 space craft signal channels simultaneously. In addition, the ARGOS system is used to locate, collect and disseminate marine environmental data from fixed or mobile DCPs.

The geostationary satellite stations fall into two classes: primary data users stations (PDUS) which receive the processed data in digital form for high resolution images, and secondary data users stations (SDUS) which receive the data in analogue form to the WEFAX standard (9). Down-link frequencies for both Meteosat PDUSs and SDUSs are either 1691 MHz or 1694.5 MHz, linear polarised. In general, the two channels lie in the range 1690 to 1697 MHz.

The PDU stations, on the whole, are much the more complex of the two. SDU stations provide images that are easily displayed on CRTs or Fax machines, using a facility known as automatic picture transmission (APT). This is also used by the polar orbiters, the image data being continuously transmitted. With the latter satellites, reception is continuous whilst it is more than about 5° above the horizon. Typically, such data can then be received over a circular area of about 2500 km radius from the antenna.

Normally, the polar orbiting satellites work in pairs separated by 90° longitude, so that each earth point is scanned every 6 hours. The TIROS-N series of satellites carry advanced very high resolution radiometers (AVHRR) whose outputs are focused on to five sensors to gather data covering the following channels:

> 0.58 to 0.68 μm
> 0.725 to 1.1 μm
> 3.58 to 3.93 μm
> 10.3 to 11.3 μm
> 11.5 to 12.5 μm

The data is digitised by an on-board computer, partly to linearise each line scan to compensate for the curvature of the earth and partly because the data will be received with fairly simple ground stations, without processing capability. After processing, the data from two channels is selected and synchronising signals that will identify the image edges are added. This composite signal is then converted back into analogue form and used to amplitude modulate a 2.4 KHz sub-carrier. This DSB AM sub-carrier, in turn, is used to frequency modulate a final RF carrier that is typically in the range 137.15 to 137.62 MHz. The transmitter power output is approximately 39 dBm (5 watts) and the transmission is right-hand circular (RHC) polarised.

APT facility is accomplished by continuous transmission at 120 lines per minute, to give an earth resolution of 4 km. The APT video format, before D/A conversion, is shown in Fig. 6.7 the receiver locking on to the required channel after identifying the appropriate sync pulse sequence. Sync for channel A is 7 cycles of 1040 Hz square wave, whilst sync B is 7 pulses of 832 pps, as explained in Fig. 6.7.

Later TIROS-N satellites also include a high-resolution picture transmission (HRPT) facility. This is a digital service that provides for an earth resolution of 1.1 km. The outputs of all five AVHRR data channels are time multiplexed (commutated) for transmission by split-phase (SP-L) modulation at a bit rate of 665.4 Kb/s on a carrier of 1698 or 1707 MHz. RHC polarisation is again used, at a level of 39 dBm.

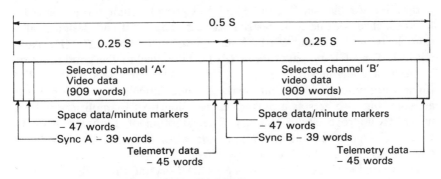

Note: 1) Equivalent data rate 41.6 kB/S
2) 10 bits per word
3) APT frame of 128 lines
4) Any two of the five AVHRR channels may be selected for transmission
5) Sync A is 7 cycles of 1040 Hz square wave
6) Sync B is 7 pulses of a 832 pps pulse train
7) Telemetry data carries calibration information.

Fig. 6.7 APT video line format in digital form.

Primary Data User Stations (PDUS)

These stations consist essentially of receiver, digital processor and image handling sections, and usually operate from antennas of 4 metres or more in diameter. This then usually involves a fairly long run of coaxial cable, between the head end electronics and the main receiver. The usual trade-off between antenna size/gain and system noise temperature is possible. The initial specification quoted in the literature (10a) calls for an antenna of 4.5 metres diameter with a gain of 33 dB at 1.7 GHz and a G/T ratio of 10.5 dB/K. This would allow the use of RF transistors with a noise temperature of about 200 K. Technical developments now allow the use of a 2 metre diameter antenna. When this is combined with a low noise head amplifier and down convertor, as exemplified by the Feedback WSR 527/535 system, an effective system can be assembled for quite a modest cost.

The receiver block diagram of Fig. 6.8 is derived from a research report prepared for the European Space Agency (ESA) by the University of Dundee (10b). In this double conversion superhet receiver, both the local oscillators work on the low side of the signal frequency. This avoids frequency inversion following the mixer stages. The first local oscillator uses a cavity-tuned microwave transistor circuit, phase-locked to the 30th harmonic of a reference, which comprises two separate crystals – one for the reception of each channel frequency. The two frequencies, 1553.9 MHz and 1557.4 MHz, produce a first IF of 137.1 MHz from the two data dissemination channels. The value of 137.1 MHz was chosen because it falls in the VHF weather satellite band and thus provides some protection from IF break-through due to terrestrial services. The front end design is also suitable to feed L band WEFAX data into an existing 137 MHz band receiver. The RF band pass filters (BPF) are designed to give good rejection of the image channel frequencies on 1416.8 MHz and 1420.3 MHz. The isolator not only helps to minimise the first local oscillator radiation by the antenna, but also acts as a buffer between the two RF amplifiers and the mixer, to improve circuit stability. The adoption of a stage of first IF amplification on the head end unit provides a convenient way of driving the main receiver, via a long coaxial cable feed.

The second local oscillator uses a 63.2 MHz crystal, the oscillator output being doubled to 126.4 MHz to produce a second IF of 10.7 MHz. This value, which gives good rejection at the second mixer image channel, is calculated from the optimum intermediate frequency formula:

$$f_{(IF1)} = (f_{(Sig)} \times f_{(IF2)})^{0.5} \qquad (6.3)$$

where, $f_{(IF1)}$ is the first IF, $f_{(Sig)}$ is the signal frequency and $f_{(IF2)}$ is the optimum second IF.

The digital PCM data is transmitted as PSK modulation, using a split-phase or bi-phase code format (a version of the Manchester code) at a bit rate of 166.67 Kb/sec; 8 bit bytes or words are organised into frames of 364 bytes, which are in turn organised into sub-frames of four or eight frames, depending on the transmission format (10a).

The first stage of demodulation uses a carrier recovery phase lock loop (PLL), which locks on to the residual carrier component in the received signal, now centred at 10.7 MHz. Phase deviations are converted into bi-polar voltage swings at the loop detector output. This is in bi-phase format,

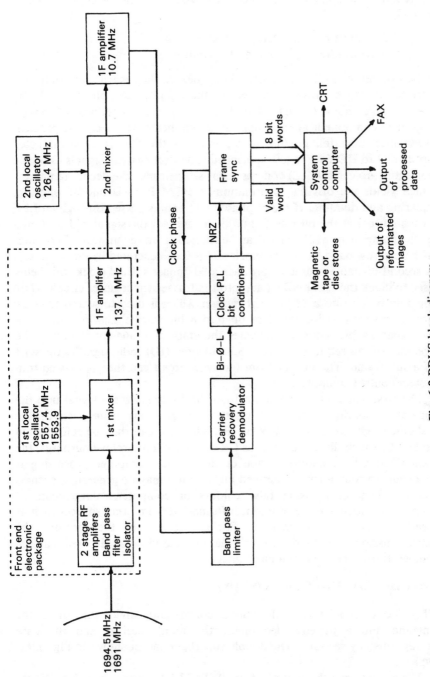

Fig. 6.8 PDUS block diagram.

where a transition at each bit cell centre is used to define the binary value as follows:

$$1 = \text{transition from 1 to 0} \ \Big\} \ \text{at bit cell}$$
$$0 = \text{transition from 0 to 1} \ \Big\} \ \text{centres.}$$

If two or more similar bits occur in sequence, then extra transitions are introduced at the trailing edges of the appropriate bit cells. These transitions, combined with the fact that the bi-phase signal does not contain a spectral component at the clock rate, complicates the decoding process. However, integrated circuits (ICs), such as the Harris Semiconductor Products Ltd HD-6408/6409, can encode and decode these signals using the non-return-to-zero (NRZ) code as a basis. Basically, the decoding process which yields the original information in PCM form is one of filtering, squaring and dividing. If the bi-phase bit stream is passed through an LPF with a cut-off at the bit rate B_r and then squared, the signal will contain a strong component at $2B_r$. This can be filtered from the unwanted products with a narrow band PLL. The clock rate is then generated by one bi-stable. A second bi-stable is used to produce a 90° phase shifted clock, this being used to clock the serial NRZ data into the frame synchroniser circuit. Here the data is run into a 32 bit shift register which is used to search for sync words and then divide the data up into 8 bit words. The presence of a particular 24 bit sequence indicates the start of a 364 byte frame. The presence of this sequence, plus an 8 bit identity (ID) code, signifies the start of a sub-frame. The output from the sync circuit can then be passed to a system control computer.

The first function of the control is to *decommutate* the signal into the separate components representing the transmissions from each of the radiometer channels (VIS, WV or IR). The data can then be reformatted and sent to such display devices as CRTs or Fax. For archive purposes, the data is stored on magnetic tape or disc. Since computer processing is involved, various forms of derived information may be generated, such as image enhancement using false colours or computer enlargements of sections of the image. In addition, information from other sources such as weather radar can be merged. Further derivations then include the retransmission of cloud motion vectors, cloud analysis, sea surface temperatures, and even animated images.

Secondary Data User Stations (SDUS)

These rather simpler ground stations, conveniently divide into three: the antenna and head end electronics, the main receiver and the data storage/display sections. The double superhet concept shown in Fig. 6.9 is typical.

The output from the space craft is 19 dBW EIRP (worst case 18 dBW) per transponder. The average free space attenuation will be in the order of 165 dB, so that the PFD at the antenna will be about -146 dBW/m^2. Antennas smaller than 2 metres in diameter can give good results, and since these have a beamwidth of about 5° to 8°, no servo-controlled steering systems will be needed.

The head end electronics consists of low-noise amplifiers and first convertor stages, a G/T ratio of 2.5 dB/K being typical. The RF section is tuned to a centre frequency of 1693 MHz with a bandwidth of 4 to 6 MHz. A crystal controlled oscillator chain producing 1556 MHz yields a first IF of 137 MHz. The WEFAX system also uses APT (the frame format is shown in Fig. 6.10), so that the main receiver section can be used with polar orbiter signals. Any image channel interference is likely to arise from terrestrial line-of-sight communications systems. For antenna elevation angles of about 5°, a rejection ratio of 80 dB will provide adequate protection from a transmitter 25 km away. This ratio can be relaxed somewhat for higher elevation angles. At angles of about 30°, 60 dB will give a similar protection (9).

The main receiver RF/IF stages can follow typical VHF/FM design, the RF stages tuning over a range 136 to 138 MHz and with a second IF bandwidth of 30 KHz. A wider bandwidth will only increase the noise level and impose a higher specification on the head end.

The second mixer stage will usually include a voltage controlled and tuned oscillator, which in the interests of frequency stability will be included in an AFC loop.

The received signal is frequency modulated, and a standard FM demodulator with a threshold of 12 dB will give satisfactory results. However, a threshold improvement of 2 to 4 dB can be useful under adverse conditions, providing a margin for non-optimum performance or slight antenna misalignment.

The demodulated signal is a 2.4 KHz sub-carrier with DSB AM. Therefore, if the index or factor of cooperation is correct, the buffered signal can be used to drive a Fax machine direct. After further decoding, the signal can be displayed on the CRT of a computer monitor or standard TV receiver.

The demodulated signal is also suitable for recording on magnetic tape. However, if an audio recorder is to be used, it is important that the record and replay tape speeds are within fairly close range agreement. It is recommended that tape-synchronous playback machines are used. The frequency of the supply driving the motor is recorded on an additional tape track and this signal recovered on playback to synchronise the motor.

A Practical SDUS

The complexity of SDU stations can vary considerably, from kit-form receiver/decoders, suitable for use as a peripheral device for a home

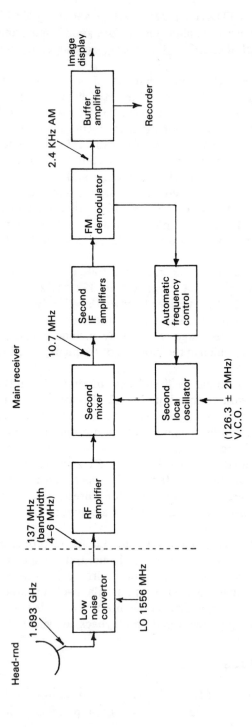

Fig. 6.9 SDUS receiver block diagram.

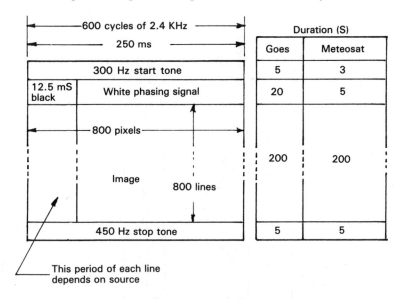

Fig. 6.10 WEFAX frame format (APT).

computer, costing around £140 (11), to the very complex systems used by national meteorological centres. The WSR range of systems produced by Feedback Systems Ltd (UK) is used in many professional, educational, private and amateur installations. These are available either as a straight VHF system or with an add-on 'S' (L) Band front end. (See Appendix A5.1.) The basic arrangement of this system is indicated by Fig. 6.11. The main VHF receiver, either processes a VHF signal from a polar orbiting satellite, or a VHF first IF signal from the 'S' Band convertor, via a switching circuit. The antennas consist of a parabolic dish for the 1693 MHz feed, or a helical or quadrifilar antenna for VHF. Both systems utilise a low-noise RF pre-amplifier mounted in the antenna. The 'S' Band first convertor is separately mounted on the antenna.

The APT receiver produces an IF of 5 MHz, which after amplification and filtering is demodulated using a phase locked loop (PLL). The 2.4 KHz sub-carrier is then decoded. The receiver has a three-frame digital memory to provide a dynamic display, which necessitates an analogue to digital conversion. At this stage, the image data is available in either digital or analogue form.

The raster timing generator is needed because the main display medium is a standard TV receiver, either monochrome or colour.

When the selected frame is read from memory, the image data is converted back into analogue form before being processed by the monochrome or colour generator stages. As the image data is basically a stepped variation of grey scale, the colour generator analyses each grey level

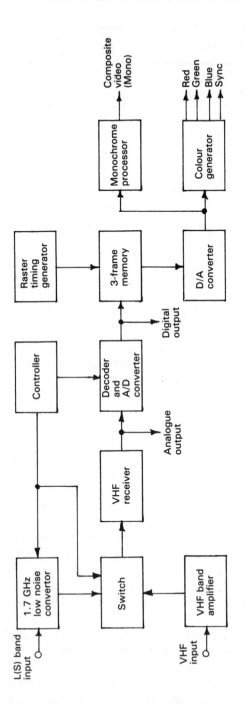

Fig. 6.11 Weather satellite receiver system (courtesy of Feedback Instruments Ltd).

and converts it into a suitable *false colour*; for instance, by displaying hot areas in red, the image details will be highlighted.

VHF Antenna Amplifier

The circuit of Fig. 6.12 shows how the MOSFET low-noise RF amplifier is coupled by a four-section tuned filter to the VHF antenna. The transistor functions as a dc-coupled, grounded source, grounded gate pair, to provide high gain with a relatively high output impedance, which is matched to low-cable impedance by the tapping on L5. The RF stage performance is optimised for use with 50 metres of cable. Where a longer run is necessary, the cable/line amplifier of Fig. 6.13 can be used. This can compensate for a further 100 metres of cable.

The APT Receiver

Figure 6.14 shows an APT receiver circuit diagram. Dc power is supplied to the head end electronics over the coaxial cable, via choke Ch1 at pad 1. The RF input signal is amplified by Tr12 and applied to gate 1 of Tr13, the multiplicative mixer stage, the first local oscillator signal being applied to gate 2. There are five separate crystal controlled oscillators, one for each channel. The selected oscillator feeds into the common frequency doubler Tr11. Since the first IF is 5 MHz, the crystal frequency can be calculated from:

$$f_x = 0.5(f_c - 5.0)\,\text{MHz}$$

where f_x and f_c are crystal and channel frequencies respectively. If alternative channels are needed, the crystal frequencies can be calculated simply. However, if the RF range is outside 136.5 to 138 MHz, the receiver section will need realignment. Tr14 together with T1, T2, and T3 provide IF amplification and filtering. Delayed and amplified AGC is applied to both Tr12 and Tr14 using IC4 and IC5.

IC3 is a high-gain limiting IF amplifier, that not only provides 5MHz output but also the AGC source and drive to the signal strength meter circuit (M1).

The IF signal is passed to IC1 of the PLL detector circuit (Fig. 6.15), where it is further down-converted to 170 KHz, to bring it within the range of the PLL detector in IC2. The recovered 2.4 KHz AM signal is passed back to the APT receiver (pads 4/13), where it is filtered and amplified by IC1. The gain of the third section of IC1 can be doubled by grounding the connection 24a & c. This is achieved via the analogue gate device IC3 on the PLL board, which shunts the feedback resistors R39 and R73 with the preset R8. This feature is needed because of the wider deviation used on the 'S' Band signals.

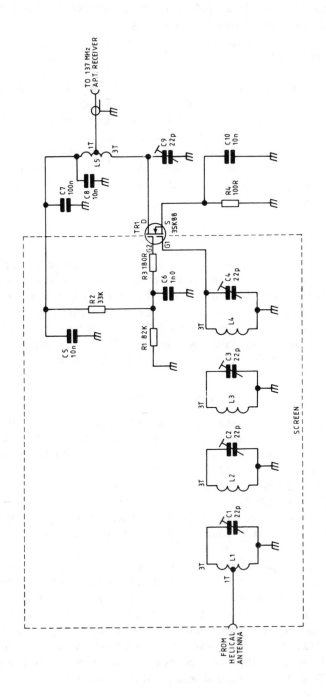

Fig. 6.12 Antenna amplifier diagram.

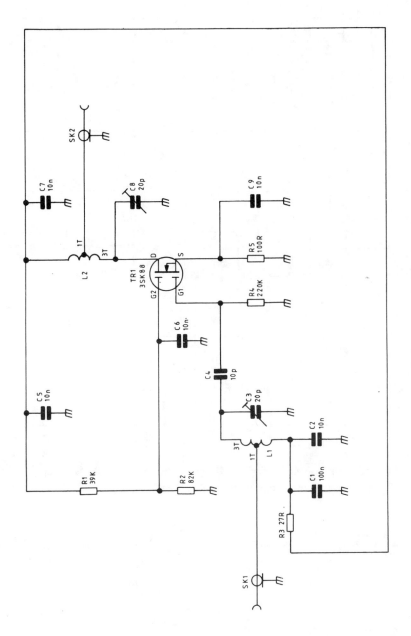

Fig. 6.13 Cable amplifier circuit diagram.

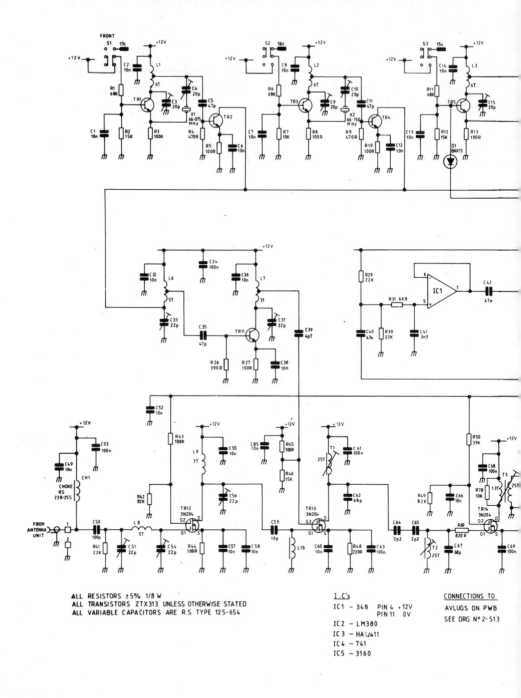

Fig. 6.14 137 MHz APT receiver circuit diagram.

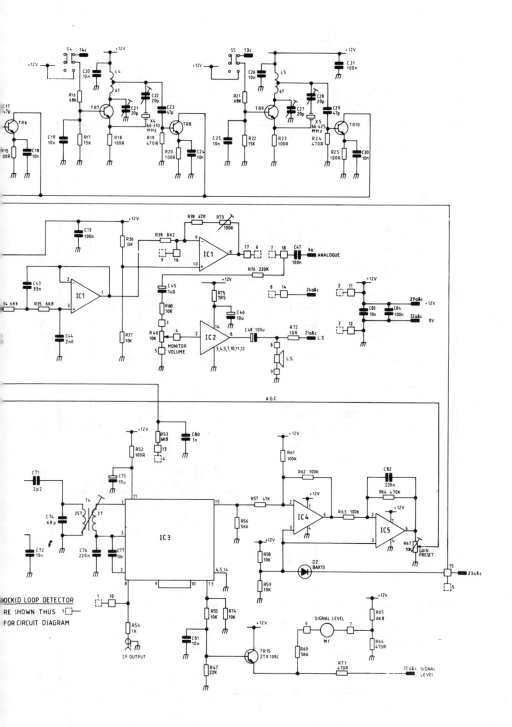

LOCKED LOOP DETECTOR
RE SHOWN THUS 1⌐
FOR CIRCUIT DIAGRAM.

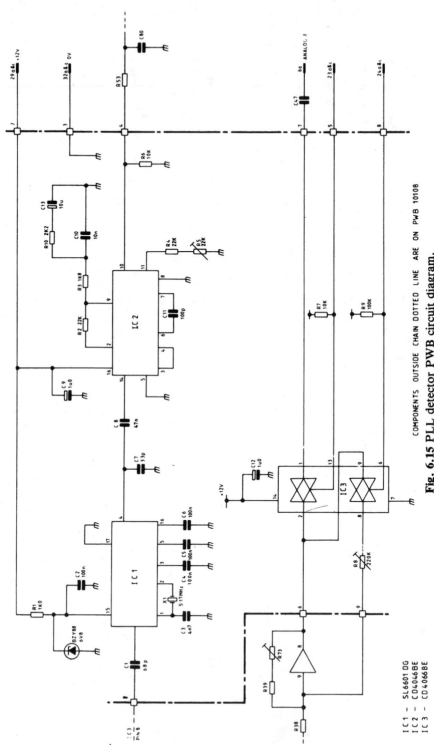

COMPONENTS OUTSIDE CHAIN DOTTED LINE ARE ON PWB 10108

Fig. 6.15 PLL detector PWB circuit diagram.

IC1 – SL6601 DG
IC2 – CD4046BE
IC3 – CD4066BE

The APT receiver may be disabled by grounding connection 23a & c, removing the AGC voltage and disabling the selection of oscillator 3, a feature also required by the 'S' band unit. An audio circuit is provided by IC2 to monitor the analogue signal.

Decoder Circuit

A decoder circuit diagram is given in Fig. 6.16. The analogue input signal is applied to the stage IC1 and then, via limiter IC2, to the PLL in IC3. The VCO in this stage runs at 4800 Hz and this is divided down by IC4 to provide two accurate phases at 2400 Hz, to generate the clock signal RECLK. R4 sets the VCO free-running frequency to 4800 Hz. The analogue signal from IC1 is fed via R7 to IC8 for amplification and then to a VCO in IC9. The output at G is a variable frequency, because the AM signal has been converted into FM. This is now decoded in an up-down counter circuit formed by IC14 and IC15. The counter counts up during the positive going period of each cycle and down during the alternative period. Therefore at the end of each cycle the counter holds a binary number that is proportional to the peak-to-peak value of the signal. This binary value is latched out via IC12 and IC13. The counter can be reset every one, two or four cycles to provide for image expansion, which is achieved via IC7. The VCO frequency is also divided in the same ratio by IC10, depending upon the setting of the pixel switches S2. IC11 provides presetting for the counter and IC16 provides the latching pulses for the output. IC16 also produces a flag signal DINSTR, to indicate that a new count has been completed. R10/IC9 controls the VCO maximum frequency so that the counter cannot overflow. R8 and R9 are used to set the maximum and minimum amplitudes, to ensure that the VCO does not cease to oscillate.

Digitisation provides 6 bit input to IC18 which converts this into TTL levels for input to IC19, the electrically programmable read only memory (EPROM). The EPROM contains data for 16 grey scale levels from the 6 bit addresses, these 16 levels being provided for by a 4 bit pattern ($W_0 - W_3$). The 6 bit address pattern is also available for further use. The grey scale range can also be preset from front panel controls.

Synchronising Signal Decoding

Synchronising signals consist of short bursts of pulses at either 256, 832 or 1040 Hz, amplitude-modulated on the sub-carrier. The actual frequency depends on the satellite. R44 taps off a part of these signals which are full-wave rectified by IC1. IC20 forms a switched, pretuned filter and comparator circuit, whose threshold is set by R30. No adjustment is provided for the 256 Hz filter, but R45 and R28 presets the 832 Hz and 1040 Hz filters respectively.

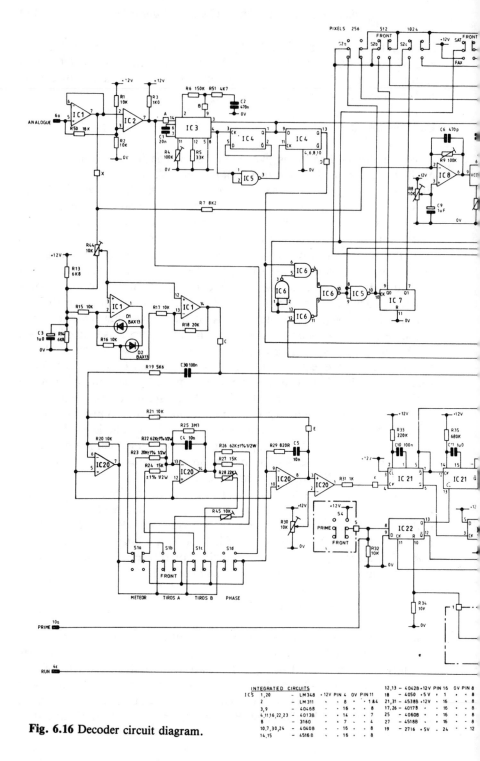

Fig. 6.16 Decoder circuit diagram.

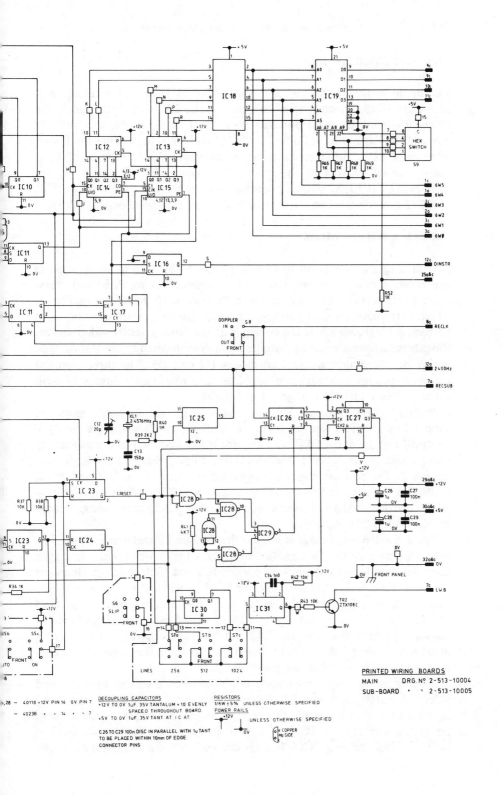

PRINTED WIRING BOARDS
MAIN   DRG. N° 2-513-10004
SUB-BOARD   "   " 2-513-10005

DECOUPLING CAPACITORS
+12V TO 0V 1uF 35V TANTALUM × 10 EVENLY
         SPACED THROUGHOUT BOARD
+5V TO 0V 1uF 35V TANT AT I C AT

C26 TO C29 100n DISC IN PARALLEL WITH 1u TANT
TO BE PLACED WITHIN 10mm OF EDGE
CONNECTOR PINS

RESISTORS
1/8W ±5% UNLESS OTHERWISE SPECIFIED
POWER RAILS
UNLESS OTHERWISE SPECIFIED

,28 — 4011B +12V PIN 14 0V PIN 7
   — 4023B "   " 14 "   " 7

Line Start Signal Generator

The main source of this signal is the 2.4576 MHz crystal oscillator and divider in IC25, which is used to generate the 2400 Hz clock signal, the frequency being trimmed by C12. This signal also provides the 2400 Hz clock in the Fax mode. IC26 and IC27 divides by 600 to produce the fastest line rate of 4 Hz. The picture slip switch changes the division ratio to 500, to allow manual phasing of the image. Line rates of 2 Hz and 1 Hz are generated by IC30, S7 selecting the required line rate. The monostable IC31 and Tr2 provide the narrow start-of-line pulses *line write begin* (LWB).

Raster Control and Memory

Circuit diagrams are given in Figs. 6.17 and 6.18. IC26 is the sync pulse generator, producing mixed syncs (MS), mixed blanking (MB) and cross-hatch (XH) outputs. This is done under the control of the master oscillator, which is trimmed by C1 to run at 2.5 MHz. Under *genlock* control, as in perhaps a TV studio where the system is used as an alternative video source, the precise frequency can be stabilised and synchronised to the station sync generator, using the two varicap diodes D2 and D5. The three identical memory sections are organised into 256×256 rows and columns of 4 bit words or pixel data ($2^4 = 16$ grey scale levels).

Reading the memory: The row and column memory addresses are supplied by counters clocked from the oscillator IC32, whose frequency is controlled by L1 and C18. This oscillator is synchronised to the master, by halting it at the end of each line; this is achieved by changing the logic state on pin 2. The row addresses are provided by counters in IC3 and IC12, whilst the column addresses are provided via IC5 and IC14. The row address strobe (RAS) and column address strobe (CAS) pulses are generated by IC37 and IC41 respectively.

The 256×256 pixel image is set into the 625 line TV raster as follows. The beginning of each field is signalled by IC49, which looks for the first sync after the end of the field blanking period. IC1 and IC10 count 256 lines per TV field, after which the data selector IC47 is switched to insert grey bars that are derived from IC3 and IC12, until the beginning of the next field blanking period. At the end of each 128th row address (beginning of the line blanking period), the counters IC12 and IC13 are stopped and preset to a number exceeding 128. When they next start to count they overflow to zero first, just as the line blanking pulse ends.

Writing to memory: A one-line buffer store is provided by IC27 and IC29, to match the large difference in data rates between the incoming signal and the raster scan. On receipt of the LWB pulse, the control circuitry associated with IC15, IC16 and IC18 goes into the *write* mode. The buffer address counter IC28/IC30 is clocked by the incoming data strobe pulses (DINSTR), divided by two. As the 4 bit words arrive, they are packed into pairs by IC20, IC24 and IC25 for storage in successive 8 bit locations in the buffer IC27/IC29. When 128 of these 8 bit words (256 pixels) have been entered, the control circuit enters the *read* mode and waits for the next field blanking period. When this starts, the buffer is read out at line rate and written into the main memory.

The main memory line address (IC5/IC14) can be slipped with respect to the raster line count (IC1/IC10). Every LWB pulse causes it to slip plus or minus one count, so that the image moves up or down the screen as the oldest line in memory is overwritten.

Depending on the direction of the satellite pass, data may be presented upside down. In order to reverse the writing process, IC46 causes an extra pulse to be added to the memory line counter (or causes one to be omitted). The lines can also be reversed from left to right by IC43/IC44.

The *busy lamp* which is controlled by IC9 and Tr3, lights on receipt of LWB pulses and goes out when the line has been loaded into memory.

Video Output: The 8 bit video output is strobed from memory by IC31 and divided into two 4 bit words by IC33 and IC36, then passed to the data selector IC47. From pads 4,6,8 and 10, these pass to the density slicer, if fitted; otherwise to the pads 5,7,9 and 11 respectively. IC47 alternates image data with grey bars derived from the address counters, if S3a is closed. S3b selects grey bars permanently.

Colour Density Slicer: The 4 bit digital video signal enters at pads 4,6,8 and 10 and then divides into two paths (Fig. 6.19).

Monochrome: The monochrome path is via IC6 and IC4, and then out as digital values to pads 5,7,9 and 11 for return to the raster control circuit of Fig. 6.17. The processed digital values are summed in the resistors R5 to R7 and the sync pulse added via R8, the circuit acting as an integrator. This provides the monochrome video output via Tr1/Tr2.

Colour generator: Comparators IC1/IC2 compare the values of the 4 bit digital video signal with values set up on hexadecimal switches. This signal is then buffered by IC5/IC7 and converted into analogue form by resistors R12 to R15 and Tr1. It now provides a common drive to each of the colour output circuits. The actual colour of each output depends also upon the digital inputs to R24, R29 and R34. These give the monochrome image false colour in certain areas. Resistors R28, R38, and R33 are used to set peak outputs, whilst R23, R27 and R32 are used to set peak white level. R37 gives control over the sync pulse amplitude.

Microwave Head Amplifier: This amplifier (Fig. 6.20) is powered from the 12 volt supply over the coaxial output cable. A 5 volt regulator IC1 is used to provide the drain supply for the gallium arsenide FET (GaAsFET) first stage, using the 5.2 volt zener diode for protection against overvoltage from a regulator failure. The negative gate supply is provided from the oscillator IC3, rectifiers D4 to D7 and the negative regulator IC2. Again, a

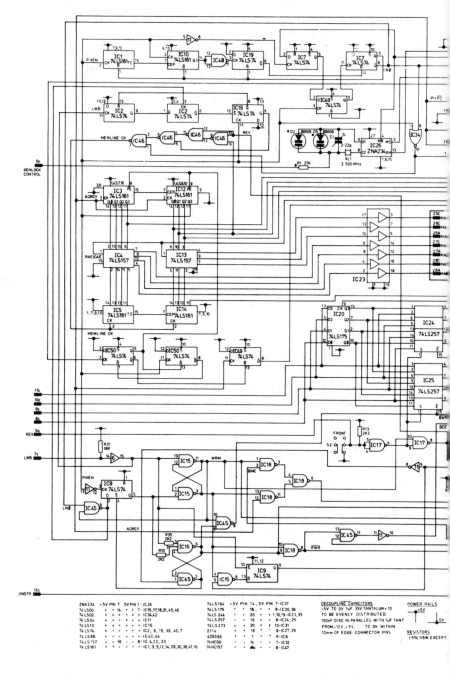

Fig. 6.17 Raster control circuit diagram.

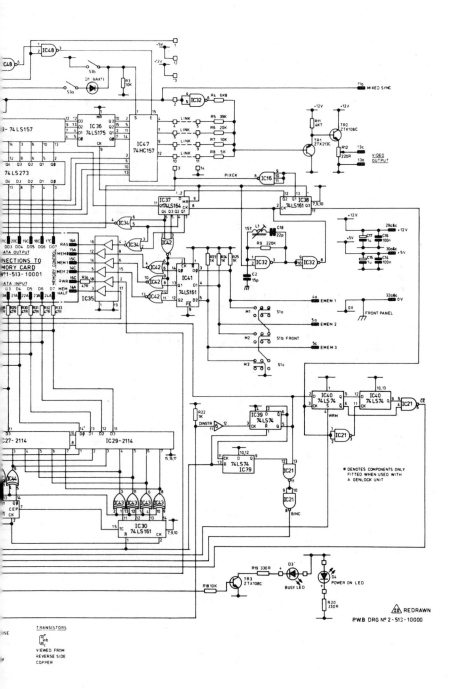

REDRAWN

PWB DRG N° 2-513-10000

TRANSISTORS

VIEWED FROM
REVERSE SIDE
COPPER

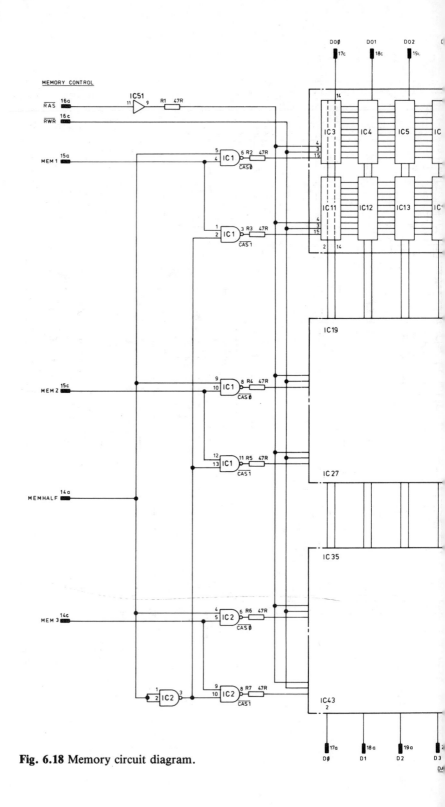

Fig. 6.18 Memory circuit diagram.

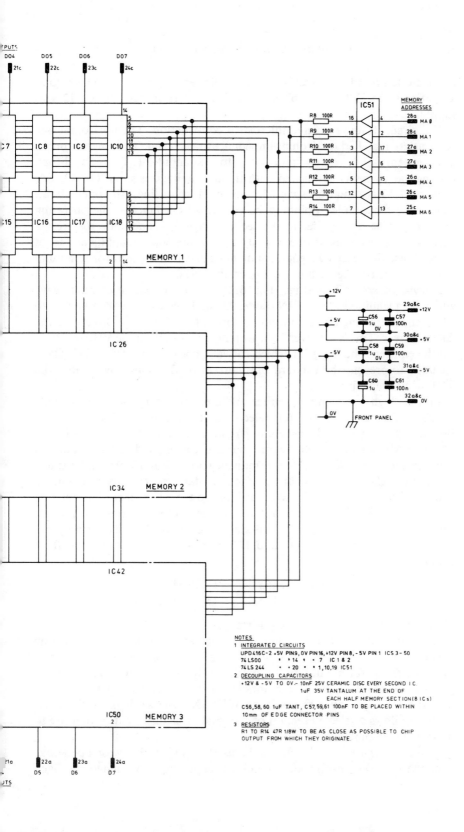

NOTES
1 INTEGRATED CIRCUITS
UPD416C-2 +5V PIN9, 0V PIN16, +12V PIN8, -5V PIN1 ICS 3-50
74 LS00   " +14 " +7   IC1 & 2
74 LS 244   " +20 " +1,10,19 IC51
2 DECOUPLING CAPACITORS
+12V & -5V TO 0V.- 10nF 25V CERAMIC DISC EVERY SECOND IC.
              1uF 35V TANTALUM AT THE END OF
                      EACH HALF MEMORY SECTION(8 IC s)
C56,58,60 1uF TANT, C57,59,61 100nF TO BE PLACED WITHIN
10mm OF EDGE CONNECTOR PINS

3 RESISTORS
R1 TO R14 47R 1/8W TO BE AS CLOSE AS POSSIBLE TO CHIP
OUTPUT FROM WHICH THEY ORIGINATE.

zener diode provides protection for possible regulator failure. The second amplifier stage Tr2 uses a bi-polar device, fed from the filtered 12 volt supply.

The circuit follows normal microstrip construction. The horizontal line sections are matched to 50 ohms impedance at each end. The antenna probe also presents a 50 ohms impedance to the input of Tr1. The broad rectangular elements associated with the decoupling resistors R1, R2, R6, R10 and R13 are quarter-wave 25 ohms sections of line. The open circuit ends cause the coupled points to be very low impedance at the operating frequency. The dc supplies to the second stage base and collector are via narrow choke sections of line. C7 adjusts the interstage coupling and is set to maximise the received signal. Input stage bias is set by R5, for minimum noise, commensurate with correct operating bias.

Down-convertor: This stage (Fig. 6.21) is a combination of microwave and VHF techniques. Two PIN diodes D5 and D6 act as series switches to either enable the high-band output or couple the VHF input directly to the APT receiver. This is achieved by a switching action from the VHF/SB line, via the emitter follower Tr11 which forward biases either D5 or D6. D4 is included to prevent the forward bias current of D6 from flowing into the VHF head amplifier.

In the high-band mode, either of the oscillator/multiplier circuits Tr1/Tr2/Tr3 or Tr4/Tr5/Tr6 is selected by the flip-flop action of the two inverters IC1. The basic oscillator frequencies are either 86.5 MHz or 86.30556 MHz, from which the third harmonic is selected. Tr7 is a common third harmonic amplifier, Tr8 is a frequency doubler and Tr9 a tripler stage. Thus the eighteenth harmonic of each crystal frequency is available to provide either 1557 or 1553.5 MHz as input to the mixer stage. The antenna input is amplified by Tr10 and filtered by the microstrip circuit to provide the second input to the mixer. This stage consists of the microstrip hybrid and two Schottky diodes D1 and D2. For the two channels, the IF is produced as follows:

$$1694.5 - 1557 \quad = 137.5 \text{ MHz}$$
$$1691 \quad - 1553.5 = 137.5 \text{ MHz}$$

Tr13 is the IF pre-amplifier output stage.

Controller: Apart from providing the necessary facilities for switching between the two modes, this circuit (Fig. 6.22) matches the 'S' Band WEFAX frame format into that suitable for the APT receiver. This receiver produces a frame format of 256 lines by 256 pixels, and this has to be inset into the 800 lines by 600 pixels of WEFAX. The settings of the BCD switches allows a selected section of the total to be displayed, giving rise to an expanded image.

Image processor: The system can be used in conjunction with a personal computer (PC). This gives the added advantages of image storage, manipulation and animation, plus the ability to provide hard copy print-outs of weather maps.

TV Studio Applications

The television broadcasters have become prime users of the satellite weather services, often using a combination of PDUS and SDUS systems plus weather information feeds from the national weather centres. In general, the digital data provided for the PDUS terminals is the more superior because of the fewer processing stages. This fact also makes this data the more up-to-date version. Since the land and sea features of each weather map that is transmitted falls on identical pixels each time, it is possible for any TV station to store its own basic maps. The locally processed images can then be overlayed on to these in a preferred *in-house* style. The weather data provided by the national weather centre will in general also include information collected from weather radar systems. The local television weather forecaster then has historic, current and predictive data on which to base a weather forecast, for either local or national broadcasting.

Environmental Satellite Services (29)

Because the data is supplied from low earth polar orbiters equipped with scanning systems, these have much in common with the meteorological services. Typically, the orbits are at an altitude of about 800 km with a period of 100 minutes. The sensors are chosen to collect data acquired by scanning the earth's surface over a range of about 500 nm to 900 nm. This represents colours ranging from green to near infra-red. High resolution images can be generated from the received data to an accuracy of less than 20 metres.

LANDSAT: This US series of satellites is operated by Earth Observation Satellite Corporation (EOSAT) and is intended to be financially self-supporting through the sale of the remote sensing information. The on-board sensors are designed to provide images that can be used for a wide range of operations. These include land use and mapping, soil erosion, crop management, volcanic activity, mineral deposits and environmental disasters. The sensors have the following earth resolutions:

> visual images, 30 metres
> cartographic images, 15 metres
> infra-red thermal images 120 metres

The data that is collected on-board is stored using high-desity magnetic tape recorders. To save wear and tear on the recorder mechanism and spacecraft power, the tape machines transmit the stored data in reverse. This avoids having to rewind and then replay. The data is relayed to the ground stations either directly on command from a control station, or via the tracking and data relay satellite system (TDRSS). For the relay of the stored data, modulated X-band carriers are used, with data rates of 15 Mbit/s for visual and IR images, and 85 Mbit/s for cartographic data.

SPOT (Système Probatoire d'Observation de la Terre): This series of French satellites is owned by a public company of which MATRA and Centre National d'Etudes Spatiale (CNES) are major shareholders. The on-board sensors provide typical environmental information with a resolution of 10 metres for colour images and 20 metres for the wide band coverage data. A unique feature involves the ability to provide 3D images from the received data. The collected data is stored on a high-density magnetic tape recorder using *three-position* modulation. A technique that provides three data bits for every two flux reversals. The relayed data is PCM encoded at 25 Mbit/s on to 20 watt carriers of either 8.025 or 8.40 GHz.

ERS-1 and ERS-2 (European Remote Sensing): This series of satellites is owned and managed by the European Space Agency (ESA) and is basically intended to provide information about the marine environment. The features being monitored include mean sea height to an accuracy within 10 cm, wave movement, sea and cloud temperatures, ice flows and icebergs, the ozone layer and climatology in general.

6.4 MOBILE SYSTEMS

INMARSAT Standard A System

One sector of commerce that was quick to adopt radio as its main means of long-distance communication was the shipping industry. The need for an easy and reliable method, not only for day-to-day operations but also in times of distress, has long been recognised. However, until the advent of satellites, the shipping services had to depend heavily on the vagaries of HF radio for long distance communications. This was later supported by VHF radio for local inshore links.

Satellites have made great changes to the services provided for mobile operation, not only at sea but also for land and air applications. Geostationary and polar orbiting satellites play complementary roles to provide these services, of which probably the best developed is that provided by the international consortium Inmarsat (International Maritime Satellite Organization) (13). This organisation provides three operational satellites in geostationary orbit and positioned at 26°W (AOR), 63°E (IOR) and 177.5° (POR). These provide a wide range of services up to about 75° of latitude. At these low angles of radiation and reception, multi-path transmissions, particularly by reflections off the sea surface, can create communication problems. Hence the need for robust signal modulation and coding techniques. The organisation virtually acts as an extension to the national post, telephone and telegraph organisations (PTTs), any service provided by a PTT being continued over the INMARSAT network.

Basically, the system is divided into three sections: a coast earth station (CES), linked to the national PTT network; a space link provided by the satellite acting as a frequency translator; and a ship earth station (SES) mounted on each mobile platform. Each CES operates with an antenna of 10 to 13 metres diameter and transmits and receives in the 6 GHz and 4 GHz bands respectively. As the Inmarsat system comprises multiple CESs in each ocean region, one has to act as network coordinator for that region. The satellite down-link to the SESs, provides an EIRP of 18 dBW, right-hand circular polarised (RHCP). The system power budget is so balanced that the mobile station can be very much smaller and less complex that a CES. Some typical features are given below.

Antenna diameter	Antenna gain	Antenna beamwidth	Noise temperature
1.2 m	23 dBi	10°	400 k
0.9 m	20 dBi	16°	200 k

The SES operates in L Band (see Appendix A5.1), all signals being RHCP, with a typical up-link EIRP of 36 dBW. The operational frequencies are as follows:

Service	Transmit (MHz)	Receive (MHz)
Maritime	1626.5 to 1645.5	1530 to 1544
Aeronautical	1645.5 to 1660.5	1545 to 1559

A critical element of the system operation is the frequency errors arising from oscillator instability and Doppler shift. To simplify the specification of each SES, AFC compensation loops are provided within each CES and in the satellites C to L and L to C Band transponders.

The services provided include telephony (companded, of frequency range 300 to 3400 Hz), digital data, analogue and digital facsimile, and electronic mail at data rates of up to 2.4 Kb/s, transmitted using FM and SCPC. Telegraphy channels provide for 50 baud (50 b/s) telex, which is transmitted in a single channel per burst mode on a TDMA carrier. High-speed computer data at 56 Kb/s is included using QPSK (4-phase). Over the space link, this uses a differential code where a logic 1 is signified by a change of state and this is followed by a 1/2 convolution coding of constraint length 7,

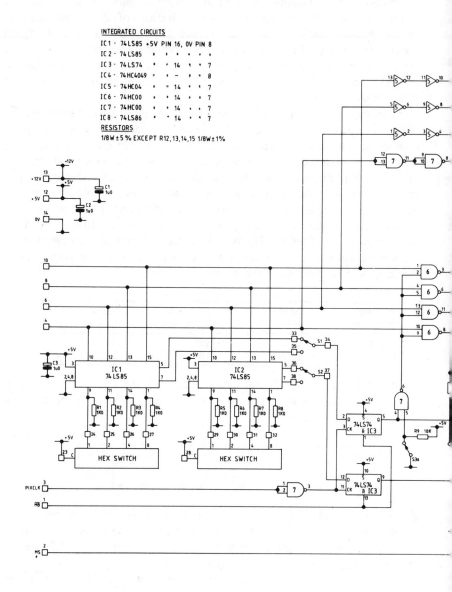

Fig. 6.19 Colour and density circuit diagram.

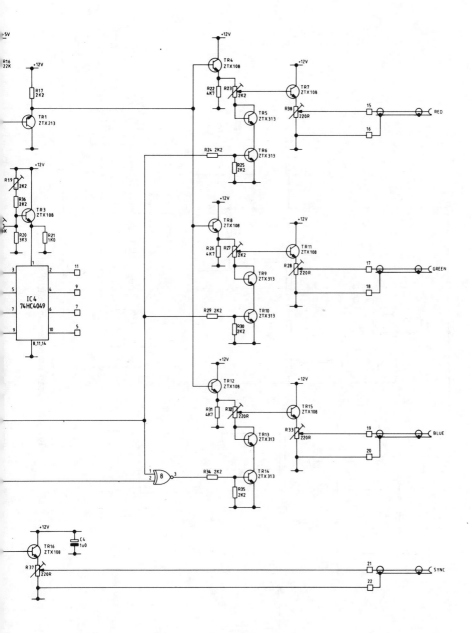

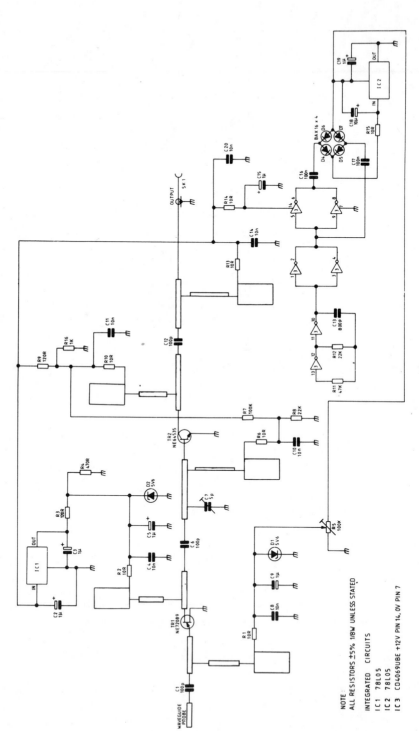

Fig. 6.20 Head amplifier circuit diagram.

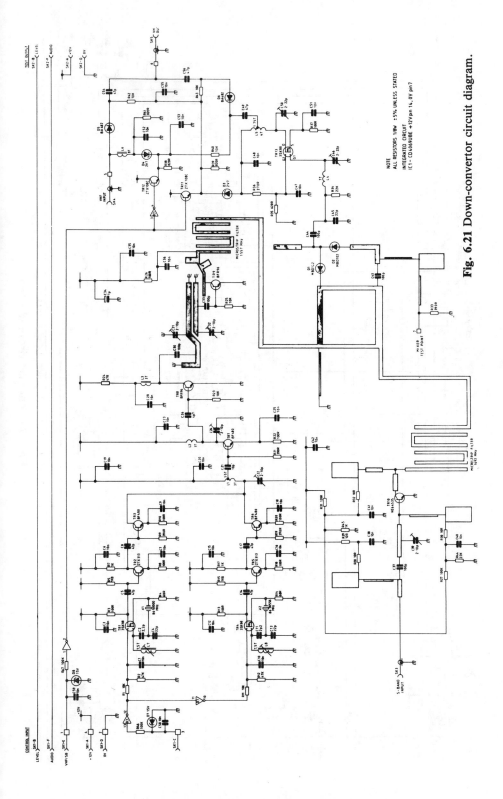

Fig. 6.21 Down-convertor circuit diagram.

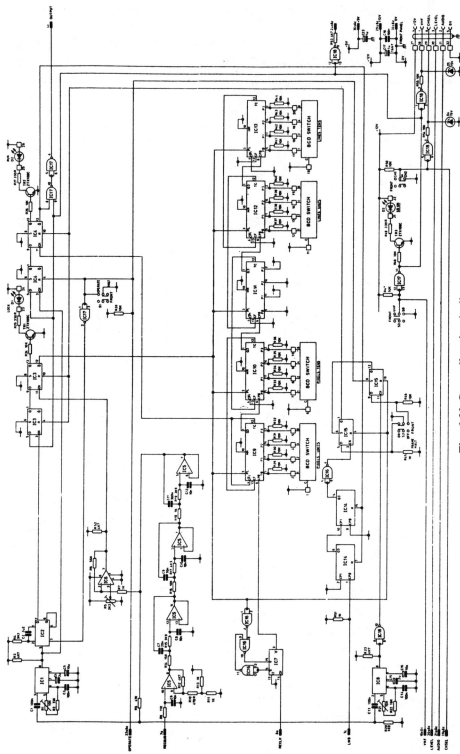

Fig. 6.22 Controller circuit diagram.

the transmitted bit rate being 112 Kb/s. In addition, other services included are meteorological and search and rescue.

Global Maritime Distress and Safety Systems (GMDSS)

The meteorological and other polar orbiting satellites often perform a dual role by providing a relay for emergency and navigation signals. These satellites are sun-synchronous in that they pass over the same point on the earth's surface at the same time each day. The operating frequencies being used are all in the VHF or low UHF bands, 121.5, 243, 401 and 406 MHz, which include the common international distress frequencies. The services are typically international, but are generally within the SARSAT/COSPAS agreements. (SARSAT = Search And Rescue SATellite; COSPAR = Cosmicheskaya Systema Poyska Avarinich Sudov, which roughly translates to 'Space System for the Search of Ships in Distress'.)

This service, which is expected to be globally operational by 1999, could be made mandatory for all sea-going vessels under the SOLAS (Safety Of Life At Sea) Convention. It already has the support of the United Nations via the work of the International Telecommunications Union (ITU) and the International Maritime Organisation (IMO), two of the organisations that have been responsible for the development of the technical standards involved.

The important emergency part of this system is based on a small beacon transmitter that, when activated either manually or automatically, continuously emits an AM signal at a power output of 100 mW. This signal is picked up by a polar orbiting satellite and repeated in real time to be received by a local user terminal (LUT) ground station. The beacon transmitters are generally known as emergency position indicating radio beacons (EPIRB) or search and rescue beacon equipment (SARBE). The position of the emergency is determined by the LUT using the Doppler shift affecting beacon signal received at the satellite. The satellite processes the beacon identification, emergency and user category codes, adds a time code and then retransmits the message.

Each LUT is capable of providing coverage over an area of about 2500 km radius and, depending upon conditions, produces a position accurate to 5 to 10 km.

A Typical Ship Earth Station

The example taken is Marconi International Marine Co. Ltd's 'Oceanray' (14). Each SES system can be subdivided into two sections, the above-deck equipment and below-deck equipment (ADE and BDE). The former comprises the 0.9 metre diameter antenna and its steering mechanism, the transmitting high power amplifier (HPA) and the receiver LNB, together with a diplexer. All are mounted within a glass fibre radome, for weather protection. The antenna gain in the transmit mode is 22 dBi. The HPA provides an output of 14 dBW or 25 watts, to give a total EIRP in the preferred direction of 36 dBW. Although these systems were initially devised for the larger ocean going ships, the relatively small size, low weight and power consumption makes such equipment also suitable for smaller vessels, such as those used by the fishing industry. A land-based version is also available, mounted on a rugged, four-wheel-drive, cross-country vehicle and designed for such things as survey operations well off the beaten track.

The antenna is initially aimed at a suitable geostationary satellite, under either manual or automatic direction control. The antenna then receives a continuously transmitted time division multiplexed (TDM) carrier from a shore station (CES). After processing, this carrier is used with an input from the ship's gyro compass, by the single board computer (SBC), to generate the necessary tracking signals so that the antenna can lock on to the chosen satellite. Compensation is provided for changes in ship direction as well as variations of attitude due to pitch and roll.

Transmission

The system operates in the *request, transmission, telex* or *voice* modes and, with additional optional circuitry, in the high speed data modes. Upon request for a call, from either the telephone or the teleprinter, the single board computer (SBC) selects the requested mode and, using data held within its memory, sets up the appropriate oscillator frequency for the channel assigned to meet the operator's requirement.

For the telex mode, data from the teleprinter is suitably formatted for transmission using TDMA. A single RF carrier may be used by up to twenty-two ships on a time-sharing basis. The TDMA time frame of 1.74 seconds is subdivided into twenty-two time slots and guard times. After formatting, the computer passes the data to the transmit IF circuit, during a time slot that has been allocated by the CES. The data is then modulated on to the selected IF carrier frequency and passed to the ADE, where it is mixed with the fourth harmonic of 365 MHz for up-conversion into L Band.

For voice mode, the signal originates at the telephone handset. The audio signal is amplified and compressed before being used to frequency modulate the allocated transmit IF carrier, which is derived from the frequency synthesiser. From here, the transmission path is the same as that for the TDMA signal.

Reception

The TDM carrier is received by the antenna and passed to the transceiver unit, where it is down-converted, again using the fourth harmonic of the 365 MHz oscillator. It is then passed to the IF demodulator circuit in the BDE. Here the signal is split into PSK-modulated or frequency-modulated components.

The PSK receiver demodulates the permanent TDM carrier and any telex signals sent to the ship, the receiver being tuned to the TDM carrier unless it is assigned to a different receive frequency by the CES. The CES thus not only allocates time slots but also receive frequencies. The SBC is used to retune the receiver, upon receipt of this information from the CES.

Voice signals are received over a channel also allocated by the CES and handled by the FM receiver. After demodulation and expansion, the audio signal is presented to the telephone handset.

The Single Board Computer (SBC)

This is based on a Z80A microprocessor together with extensive memory capability. This consists of random access memory (RAM), which is battery supported in the event of power failure, programmable read only memory (PROM) and electrically erasable programmable read only memory (EEPROM). Apart from the overall management of the system, the SBC acts as a buffer for telex data. This is received at either 4.8 or 1.2 Kb/s and can only be input to the teleprinter at 50 bauds. The RAM section of memory allocated to this function can hold up to 16,000 characters.

The computer also has a built in self-test routine for use on power-up or after power failure situations. This includes:

- Writing data into RAM, which is then read and compared with the original to check this section of memory.
- Checking the state of the backup battery and the real time clock circuits to signal the period of power-down.
- Using the power-down time to initiate the antenna search sequence to relocate the satellite.
- Checking the data held in EEPROM.
- Checking the status of its input/output ports.

Above Deck Equipment (ADE)

Antenna pointing is controlled and stabilised against pitch and roll by gyroscope motors, and in azimuth and elevation by control signals provided by the SBC, dependent upon its various inputs. The current azimuth and elevation angle analogue information is derived from two potentiometers. This is converted into digital form for local processing and then sent to the SBC as serial data over the TX IF IN line. Control data to the platform is received in serial form over the 365 MHz IN line, the low-frequency data signals being separated from the high frequencies with filters. (See Fig. 6.23).

The gyroscope motor platform is normally clamped to the antenna by solenoid action. This can be released when it becomes necessary to move the antenna independently.

The HPA in the transceiver section is only energised in the transmit mode. The diplexer directs the received signal to the down-convertor or the transmit signal to the antenna feed horn. This section also generates the

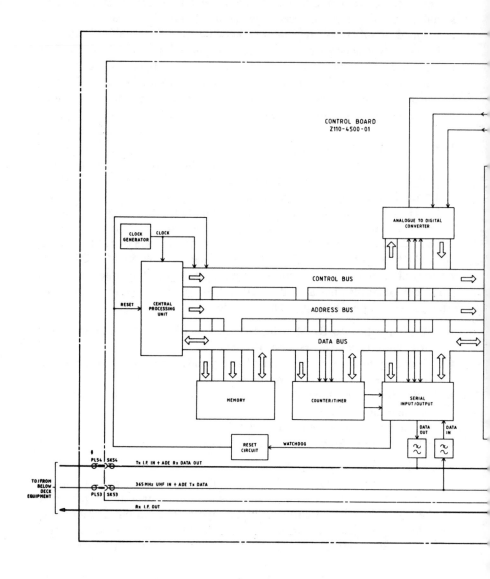

Fig. 6.23 Above-deck equipment.

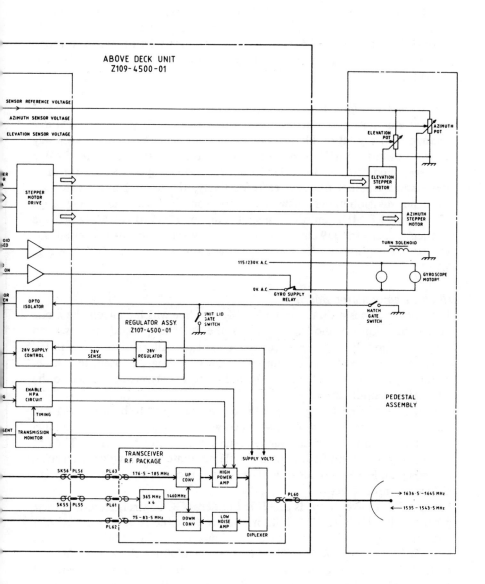

ABOVE DECK UNIT
Z109-4500-01

SENSOR REFERENCE VOLTAGE

AZIMUTH SENSOR VOLTAGE

ELEVATION SENSOR VOLTAGE

AZIMUTH POT

ELEVATION POT

STEPPER MOTOR DRIVE

ELEVATION STEPPER MOTOR

AZIMUTH STEPPER MOTOR

TURN SOLENOID

115/230V. A.C.

0V. A.C.

GYRO SUPPLY RELAY

GYROSCOPE MOTORS

OPTO ISOLATOR

UNIT LID GATE SWITCH

HATCH GATE SWITCH

REGULATOR ASSY
Z107-4500-01

28V SUPPLY CONTROL

28V SENSE

28V REGULATOR

ENABLE HPA CIRCUIT

PEDESTAL ASSEMBLY

TIMING

TRANSMISSION MONITOR

TRANSCEIVER
R.F. PACKAGE

SUPPLY VOLTS

SK56   PL56   PL63   176·5 – 185 MHz

UP CONV

HIGH POWER AMP

PL60

1636·5 – 1645 MHz

365 MHz × 4   1460MHz

SK55   PL55   PL61

1535 – 1543·5 MHz

75 – 83·5 MHz

PL62

DOWN CONV

LOW NOISE AMP

DIPLEXER

fourth harmonic of 365 MHz (1460 MHz) to provide one input to each of the two mixer stages. The incoming frequency range of 1535 to 1543.5 MHz is converted into 75 to 83.5 MHz by subtraction, whilst the transmit range of 1636.5 to 1645 MHz is achieved by addition (176.5 to 185 MHz + 1460 MHz).

Below Deck Equipment (BDE): Synthesizer Board

The system master frequency is provided from a 5 MHz temperature-compensated crystal oscillator. The output can be compared for accuracy using an external 5 MHz standard. The two outputs at 1 MHz and 5 MHz are used to control the 365 MHz local oscillator and generate the transmit and receive channel frequencies respectively, the latter being created by frequency division under software control from the SBC. The receive range of 64.3 to 72.8 MHz and the transmit range of 176.5 to 185 MHz provide frequency space for 339 channels of 25 KHz. Various out-of-lock situations are signalled, both by LEDs and over the data bus, to the SBC. (See Fig. 6.24.)

IF and Demodulator Board

The SBC provides a disable/enable PSK/$\overline{\text{FM}}$ line which controls the mode of transmission. Audio signals from the interface board are used to frequency-modulate the selected Tx IF carrier. Alternatively, formatted data from the SBC, after amplification/encoding, is used to phase-modulate the carrier. Whichever signal has been generated is passed to the ADE after amplification.

The received first IF signal is split to provide inputs to both the FM and the PSK receivers. Two mixers, supplied with the necessary receive channel frequencies, from the synthesiser generate second IFs of 10.7 MHz. The FM channel signal is amplified, limited and demodulated to provide the received audio signal. In the absence of an FM carrier, the audio channel is muted to reduce the noise level.

The PSK receiver input is amplified under AGC control, from a voltage generated by the PSK demodulator. Due to this, the IF level is practically constant under a wide range of input signal levels. In addition, the PSK modulator also provides the TDM data and a 1200 Hz clock which is used by the SBC. The AGC voltage is also used by the SBC as an indicator of signal level to detect the correct alignment of the antenna. Any fault conditions are monitored by the SBC and also signalled to the operator by LEDs.

Interface Board

This unit forms the communications link between the transmitter/receivers, the SBC and the communicating peripheral devices. These include the gyro compass, the telephone/PABX and the automatic alarm/distress systems.

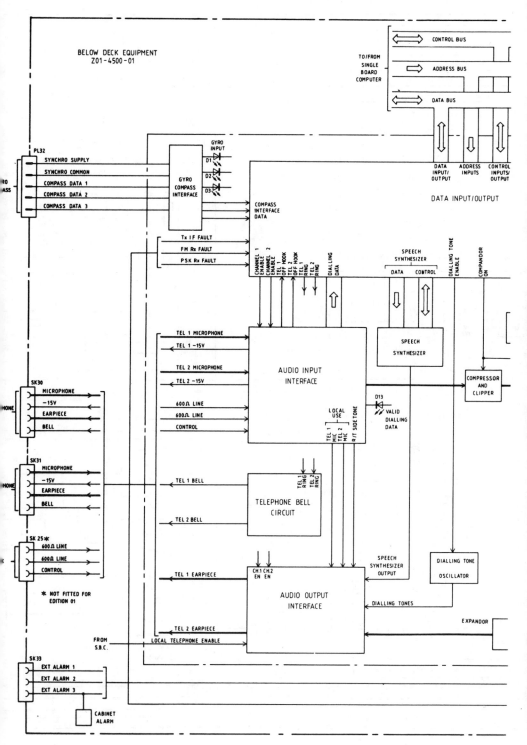

Fig. 6.24 Below-deck equipment.

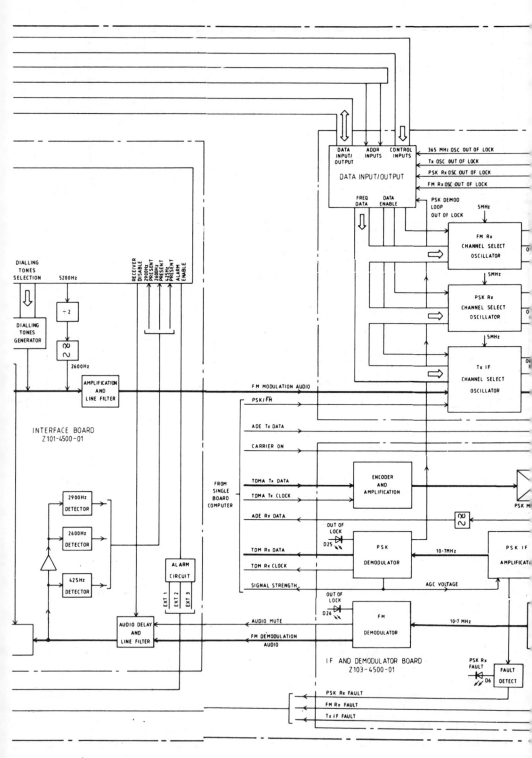

Fig. 6.24 *Continued*

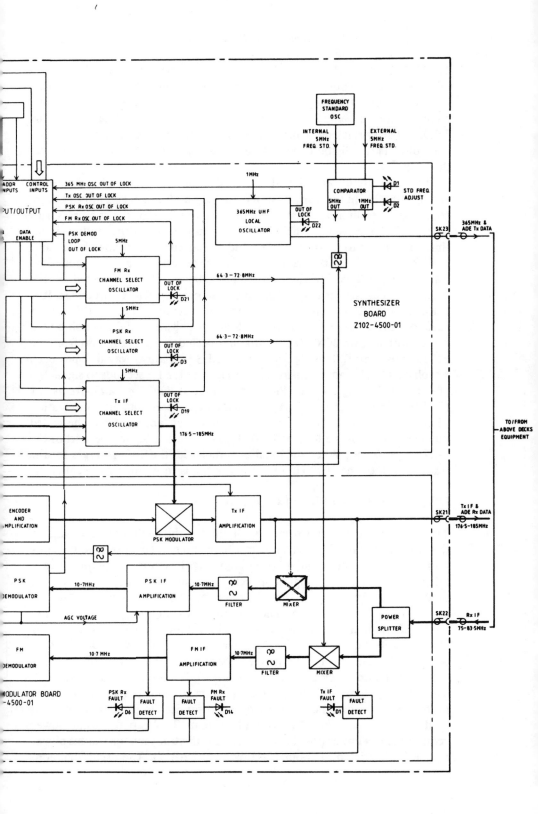

The section also includes the generators for the necessary signalling tones and the speech synthesiser. This latter provides *user friendly* synthesised instructions over the telephone, for even for the inexperienced operator, the audio responses being obtained by the SBC from word data held in the PROM. The synthesiser strings these together to form the necessary phrases and converts the digital codes into an analogue signal to be delivered over the telephone.

The audio circuits are filtered in each direction to ensure a bandwidth of 300 to 3400 Hz. In the transmission mode, the audio signal is clipped before compression to limit any transients that are too fast for the compressor to react to. The receive path includes a short delay to allow the muting circuit time to respond to the FM carrier, and a complimentary expander circuit.

When the ship makes a request for transmission, a CES replies by transmitting a 2600 Hz tone on its FM carrier. This is detected by the receiver and causes the channel select oscillator to change, to produce a 10.7 MHz IF with 2600 Hz tone. After demodulation and filtering, this tone is used to signal to the SBC that the CES has established a link. The ship then replies with its own 2600 Hz tone that has been derived by division from the 5200 Hz oscillator. This signifies to the CES that a two-way link has been established, and it now transmits a 425 Hz tone for 1.5 seconds to signal 'go ahead with dialling'. The dialling system used is *dual tone multi-frequency* (DTMF), where each digit is represented by a different unique combination of two audio tones. When the CES receives an answer from the called number, it transmits a tone of 2900 Hz for 150 mS to signal the start of the call charge period. When the handset is replaced, the SBC arranges for the transmission of the 2600 Hz tone for 2 seconds to terminate the call.

INMARSAT Standard B and C systems

The Standard A system is somewhat limited in the facilities offered and to the size of the vessel on which it can be installed. To overcome some of these problems, the Standard B system was developed. This provides a telephony service using digital speech processing and coding plus QPSK modulation. One of two digital communications data rates (9.6 and 16 kbit/s) are user selective. The system uses adaptive predictive coding with FEC (forward error correction) and a 1/2 rate convolution code, plus Viterbi soft decoding at the receiver. The telephony mode is voice operated to allow faster and simpler operation of this lower cost SES.

The Standard C system shown in Fig. 6.25 (Courtesy of Thrane & Thrane Ltd, Denmark), uses the same transmission frequencies as Standard A and was designed to extend digital services to smaller vessels and land vehicles over the INMARSAT system. To keep the total power requirement as low as possible required that the transmission output power should be restricted to about 15–20 watts (13 dBW). The usual Shannon-

Fig. 6.25 INMARSAT-C receiver system (courtesy of Thrane and Thrane Ltd, Denmark).

Nyquist S/N ratio trade-off thus implies that operation at a low C/N ratio, means that the system has to operate at a low bit rate and a narrow bandwidth. The resulting design is a low cost, flexible, go anywhere, satellite communications system that is capable of interconnecting with a mobile terminal to practically any PSTN system.

The basic data rate is 600 bit/s, but this can be doubled in the redundant mode, when each message is sent twice. The coding method uses a 1/2 rate Viterbi code which can be interleaved for added error protection to give a bit error rate (BER) better than 10^{-3}. The modulation of the RF carriers is by binary PSK (BPSK). The system frequency tuning is synthesiser control in steps of 5 KHz. As indicated by Fig. 6.25, the mobile unit uses a non-stablised, omni-directional antenna and the specification calls for G/T figure of merit of -22.8 dB/K. The transmission coding and system protocol is based on the CCITT X25/X400 standards, with the data organised into packets with 8.64 second duration frames.

The service also provides facilities for the mobile terminal to transmit position and data reports. These consist of up to 32 bytes contained within three packets. Both position and data reports can be initiated either by polling or interrogation from a base station. Poll commands instruct the terminal how and when it should respond. There are three types of polling signals available, for individual, group or area responses. In the last two cases the polling terminal arranges for queueing control. An *enhanced group call* service is also available for messages fed into the Standard C system via a terrestrial PTT network. This allows messages to be routed to mobiles according to a priority. A single message can be received simultaneously by a number of mobiles anywhere within global range, or by mobiles just within a specific region.

The FleetNET system which operates over INMARSAT C, allows individual companies to communicate with all their associated mobiles in a broadcast manner, or individually by using a unique addressing technique. In addition, navigation sensors can be added to the mobile terminals. This allows them automatically to transmit regular position reports to the home base station.

SafetyNET is exclusively available to the maritime user. This is used for the transmission of safety related information such as weather warnings for specific areas.

Since the Standard C terminals require very little power and can operate in an automatic mode, they are sometimes used for the remote sensing and reporting of terrestrial situations. This supervisory control and data acquisition (SCADA) mode allows the terminals to be positioned to gather data from such situations as volcanic, geothermal, pressure ridges and plate movement, thus providing advanced warning of developing environmental disasters.

INMARSAT Aeronautical Mobile System

This developing service operates by using the satellite as a relay link between the aircraft and a ground station, which in turn provides the interconnection to the terrestrial PSTN system. The service, which operates in L(S) band, provides both passengers and aircrew with an almost global voice and data communications link. Two digital voice channels are provided at 9.6 kbit/s for passenger use, a single channel at 4.8 kbit/s for crew use, and 600 bit/s for a digital data messaging service. Due to the high aircraft speeds, Doppler shift could be a problem. The RF, modulation and coding schemes are devised so that provided the Doppler shift does not exceed $\pm$ 100 Hz, the system can provide a high quality of service. The aircraft antennas, which are encased in an aerofoil shaped housing, are mounted externally and take on the appearance of *shark fins*. To provide all-round coverage and maintain constant contact with the satellite, the antennas are of the steerable, planar, phased array types. The voice signal is PCM coded by sampling at 8 kHz to produce 8 bit samples every 125 μs. These are processed by a microprocessor and the bits are clocked out at 1536 kHz. The 8 bit bursts at 1536 kHz are then converted into a continuous stream at 9.6 kbit/s. This output is then multiplexed with a dual tone multi-frequency dialling tone, to generate a gross bit rate of 10.368 kbit/s. After the addition of FEC bits to counter the effects of multipath fading, the bit rate has doubled to 21 k/bits. This serial bit stream is then modulated QPSK on to the final L(S) band carrier.

Ultimately the system is expected to be allocated a wide enough frequency spectrum to accommodate engineering, navigational and other data. This could be used to improve the safe and economic operation of the aircraft.

PRODAT Data Service

The European Space Agency (ESA) is developing a system known as PROSAT, to be available for land, air and maritime mobile operations. This is referred to as PROMAR. A low cost digital data only service that is included in the scheme is known as PRODAT.

This system operates through existing ESA ground stations and IN-MARSAT communications satellites and provides a low-speed, low-power and narrow bandwidth service. It is initially intended to work alongside the more common HF communications radio systems currently used by aircraft. Since this mode suffers from the vagaries of variable propogation, the satellite system will provide a more consistent and reliable service. The flexible system design allows for communications mobile to ground, mobile to mobile, and ground to mobile. In addition, it provides for extension to both land and maritime operations.

The ground station transmits to the satellite in the 6 GHz band using time division multiple access (TDMA) with differential binary phase shift keying (DBPSK). Each TDM frame is 1.024 seconds long and is divided into 64 bit slots. Each slot representing a channel sending data to a user at 47 bit/s. The other bits in each slot are used for coding and control. This signal is then relayed to the mobile by a transponder of 24 dBW output using a 1.5 GHz link. A lower power output is envisaged by using fewer slots per frame. PRODAT uses a two-dimensional Reed-Solomon (RS) interleaved code that is capable of correcting both random and burst errors. In addition, if the burst errors are too long for correction, an automatic request repeat (ARQ) for retransmission can be generated, acknowledgement of accurate reception being given during the transmission of the following block. Each block is numbered due to the delay that occurs over the satellite link.

Quadrifilar helix antennas with 0 dBi gain and a hemi-spherical radiation pattern, are mounted externally to the aircraft and housed within a low drag aerofoil casing.

The aircraft transmissions use the 1.6 GHz band to the satellite, which in turn relays the signal to ground using a 6 GHz carrier. The down link uses code division multiple access (CDMA) with suitable collision detection. The *spread spectrum* signal is generated by the addition of a PRBS (pseudo random binary sequence) to the channel signal. The system allows for 32 simultaneous channels using the same carrier frequency and occupying a total bandwidth of 650 kHz. The CDMA signal is decoded at the ground station using correlation detection and the appropriate PRBS code, before being directed to the user terminal.

6.5 SATELLITE NAVIGATION SYSTEMS

Transit System

Specialised receivers tuned to the transmissions of polar orbiting satellites of the US Navy Navigational Satellite System (alternatively known as the 'Transit System') can provide very accurate navigational positions or fixes. Although the system was originally intended for military purposes it is, with certain restrictions, available for civil applications by both water and land-based mobile vehicles.

The space link part of this system consists typically of five satellites, at an altitude of about 1100 km, in polar orbits, with periods of approximately 107 minutes. The RF radiated power is of the order of 1 watt. The transmitted data includes a time code and precise details of the orbit parameters, a fix being obtained by decoding the data and measuring the Doppler shift in the received carrier frequency. The information is then evaluated in a computer controlled system, in the manner depicted by Fig. 6.26.

The satellite's precise orbit is not circular, but affected by minor perturbations. These would lead to errors in the calculated positions. To ensure that the data is as accurate as possible, ground control stations

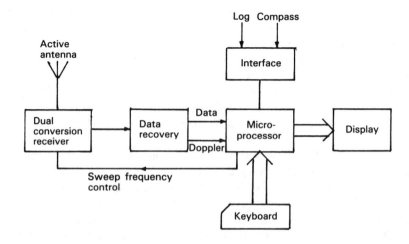

Fig. 6.26 'Transit' Sat-Nav receiver systems.

monitor each satellite transmission and use a computer to establish the true orbit. Updated, corrected information is then transmitted back to the satellite, twice in each 24 hour period.

A satellite continuously transmits messages of precisely 2 minutes duration on frequencies of 150 and 400 MHz, the latter being commonly used for civil navigation purposes. Each message consists of 156 words of 39 bits plus one 19 bit word, making a total of 6103 bits at an equivalent rate of 50.858 b/s. The first three words define universal time (GMT) and the next 25 words, used for navigation, define the minor perturbations and the fixed parameters of the current and future orbits. The remaining words are for military purposes and are not decoded. The primary bit format is shown in Fig. 6.27(a), with a duration of 19.7 mS. The bit pattern is used to

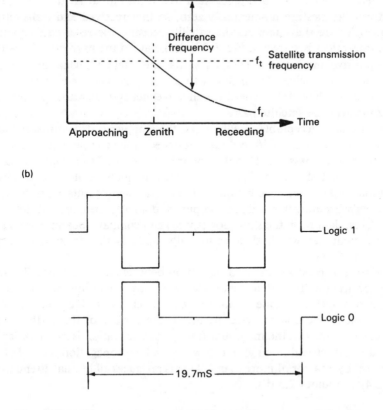

(a)

(b)

Fig. 6.27 (a) Binary logic format. (b) Doppler frequency relationships.

provide PSK of $\pm 60°$ of the 399.968 MHz carrier. This form of modulation is chosen so that the Doppler shift in the received carrier frequency can be easily and continuously measured.

The received frequency will be different to that transmitted due to the Doppler effect, which depends upon the relative velocities of the satellite, the receiver and the axial velocity of the earth. A simple mathematical formula for the observed change of frequency is derived in the literature (15), Showing that the Doppler shift is given by;

$$f_t - f_r = f_t(2v/c) \tag{6.4}$$

where:

f_t = transmitted frequency
f_r = received frequency
v = relative velocity of source and receiver
c = 3×10^8 m/S.

As the satellite rises towards its zenith, approaching the receiver, the received frequency will be higher than that transmitted. At the zenith the relative velocities are momentarily zero, so that received and transmitted frequencies are the same. As the satellite recedes, the received frequency falls; this effect is shown in Fig. 6.27(a). To avoid the problem of dealing with positive and negative Doppler shifts, the satellite navigation receiver compares the received frequency with a highly accurate and stable local frequency of 400 MHz. The difference frequency thus varies around 32 KHz during each satellite pass.

During each acceptable satellite pass, the Doppler shift is measured and counted over periods of 30 seconds and stored in the computer memory. By the end of each pass, the Doppler count has been used to compute a curve similar to that shown in Fig. 6.27(a). This uniquely relates the receiver position to the orbit. The computer then uses this, together with the data from each 2 minute message, to compute and display the position. Between such fixes, data from the ship's log (speed) and compass (bearing) are input to the computer which then continually updates the position by dead reckoning (DR).

Orbits that produce angles of elevation between about 15° and 75° can produce accurate fixes that even under worst case conditions can be within 500 metres. Orbits outside of this range are likely to produce unacceptable errors due to the non-spherical nature of the surface of the earth. At an altitude of 1100 km, the minimum free space attenuation is in the order of 145 dB. At elevation angles below 15°, the propagation path length increases by a factor of more than 2.4. This produces additional attenuation of $(2.4)^2$, or about 7.5 dB.

TRANSIT Sat-Nav Receivers

Figures 6.28 and 6.29 show the block diagrams of a typical satellite navigation receiver, the Walker Marine Sat-Nav 412 (16, 17). The later versions, the Navstar 603S and A300S, follow similar principles but make more extensive use of surface mounted component and VLSI technology. Additional inputs are also provided for interfacing to other navigational system receivers, auto pilots and printers.

The RF/IF sections, shown in Fig. 6.28, consist of an active antenna, using a low-noise RF amplifier to provide a gain of 16 dB with a bandwidth of 10 MHz at 400 MHz, power being supplied over the coaxial feeder cable. A second RF amplifier is used at the input to the main receiver, this being a dual gate MOSFET device. Apart from reducing local oscillator radiation and rejecting out of band frequencies, this second RF stage allows the system to function, with somewhat reduced efficiency, in the event of storm damage impairing the antenna performance.

The main receiver is a double conversion superhet, designed to handle what are effectively narrow band frequency modulated (NBFM) signals. The two local oscillator frequencies are derived from a highly stable, temperature-controlled master crystal oscillator (2 parts in 10^{10}), using frequency multipliers. Local oscillator frequencies of 375 MHz and 25 MHz produce first and second IFs of 24.968 MHz and 32 KHz respectively. The first IF stage contains a monolithic crystal filter and a high-gain, limiter-type integrated circuit, to ensure a constant level of signal for the following signal processing stages. The second mixer produces an IF output of 32 KHz which carries the Doppler shift component, nominally of about ± 8 KHz maximum. By the use of strip-line and low-noise techniques, these receivers achieve a sensitivity of about 0.3 μV Rms and a noise factor of less than 3 dB.

The signal processing and data recovery action can be described using Fig. 6.29. The nominal IF input of 32 KHz tracks across the band, 24 to 40 KHz due to the Doppler shift in the 'Transit' signal. The phase-modulated data and bit sync is extracted by phase demodulation and rectification, the demodulator switches being driven by the quadrature signals I and Q. The quadrature phase generator clock inputs are driven from the phase locked loop (PLL) voltage controlled oscillator (VCO), which runs at four times the tracking frequency. This is then divided by four in the two D-type flip-flops to produce a frequency 'f' of two square waves in quadrature; f tracks, either during a forced ramp mode or under satellite lock, between the frequencies of 15 and 42 KHz. During a period of 'no signal', the loop is tracking in a fast forced ramp mode, the search mode; the receiver noise is demodulated and this is averaged out to zero by the data filters.

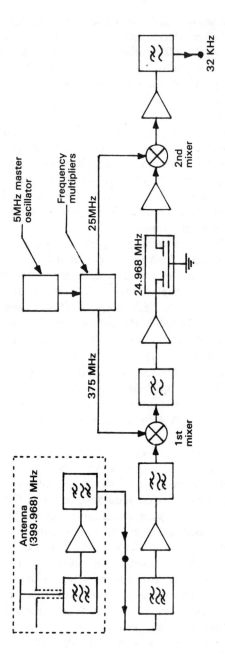

Fig. 6.28 Sat-Nav receiver for signal processing.

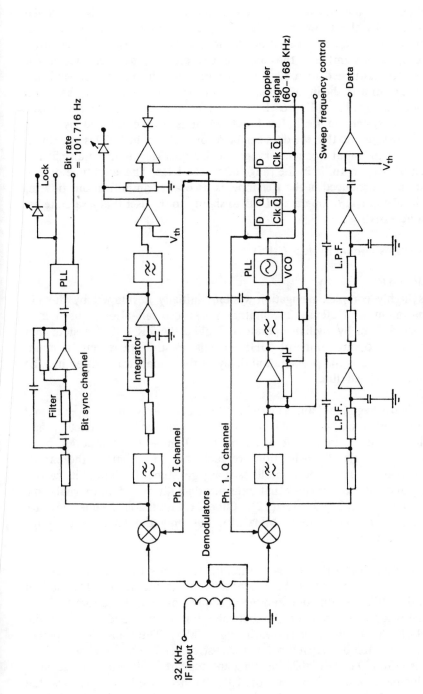

Fig 6.29 Data recovery demodulators.

When a coherent signal is present at the IF input, it is effectively multiplied by the I switching signal. After filtering, this produces a dc component plus a ripple signal in the bit sync channel. The presence of the dc component signifies to the computer that a signal has been detected and the PLL is switched to a slow ramp mode, which allows it to pull into synchronism with the input signal. Once this is achieved, the dc offset ripple becomes a signal at 101.716 Hz that is used for data clock synchronism. With the system in lock, the input signal is also multiplied by the Q switching signal so that the data can be recovered. As the VCO is tracking at four times the input frequency range, the Doppler shift is also multiplied by a factor of four, to lie in the range 60 to 168 KHz. The amplitude of a dc component produced in the I channel is used after amplification, filtering and level control, to operate a LED to signify to the user that a satellite has been accessed.

Global Positioning System (GPS)

NAVSTAR
This highly complex navigation system, initially developed by the US Department of Defense for military purposes, can also be used in a restricted form by civilian operators. Unlike the TRANSIT system that uses the Doppler shift of UHF signals as the measuring domain, NAVSTAR uses the propogation delay of the signals transmitted in L(S) band by each satellite.

The Space Segment (30): This will ultimately consist of 18 operational satellites plus six spares. These will be positioned in six circular and inclined orbits at an altitude of about 20,170 km and with a period of 12 hours. The angles of inclination are such that any point on the earth's surface can see at least four satellites at any give time. Each satellite carries a very accurate atomic clock that transmits digitally-coded time signals on a frequency of 1227.60 MHz(L2). Other transmission frequencies are 1381.05 MHz and 1575.42 MHz(L1), but only the latter is used for navigation purposes.

The Coding System (31): Each satellite transmits on the same carrier frequency of 1575.42 MHz but is identified by its unique GOLD code that provides BPSK modulation and direct sequence-spread spectrum (DS-SS) transmission. The relationship between this carrier and those of the INMARSAT-C system is shown in Fig. 6.30(a). The transmission power level is at −163 dBW, well below the system general noise level.

Each GOLD code is 1023 bits long and repeated 1000 times per second, producing a code chip rate of 1.023 Mbit/s. Modulation raises the nominally 100 Hz wide base band up to a transmission bandwidth of about

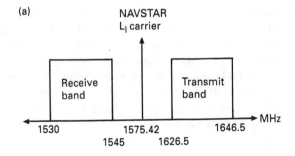

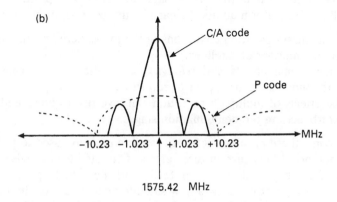

Fig. 6.30(a) NAVSTAR L_I frequency relative to INMARSAT-C frequencies. (b) Spectra of NAVSTAR C/A and P codes.

2 MHz. This code, referred to as the coarse acquisition code (C/A), provides the standard positioning service (SPS) for civilian applications.

A second code, referred to as the P code, at a rate of 10.23 Mbit/s is available for military purposes and provides the precise positioning service (PPS). This code is about 6 Mbits long and would repeat every 267 days. However, this sequence is reset weekly and with a different starting point for each satellite. The spectra for the two codes are shown in Fig. 6.30(b). As an indication of the accuracy of positioning, 1 bit of the SPS code is equivalent to 300 metres, whilst that for the PPS code is about 1/10 of this. Significantly better accuracy can be achieved with both codes by using a phase measuring technique in the receiver.

The Command Segment: Because of the variations in earth geometry and the vagaries of propagation, each satellite additionally transmits data about the parameters of the present and future orbits of all the satellites. This *almanac* allows the receivers to calculate accurate positional fixes and to be prepared with the correct code as a new satellite appears over the horizon. Information for the almanac is collected by ground stations for processing in the system central computer at the master control station in Colorado, USA. Four other ground stations located in Ascension Island, Diego Garcia, Kwagale and Hawaii ensure that any satellite is never out of contact with a ground station for more than two hours, the regular updating period. To allow a significant margin of error the almanac remains valid for up to four hours.

The User Segment: In order to obtain a fix in three dimensions it is necessary to obtain data from three satellites. However, positional accuracy will not be uniform across the earth's surface, due in the main to:

(1) Changing propagation conditions in the ionosphere and the use of a minimal number of satellites.
(2) Gravitational effects and *relativity*. The latter feature causes the earthbound clocks to appear to run slow.
(3) The effects of multipath reception can give rise to noise and phase disturbances in the demodulation stages.

To overcome these problems a fourth satellite is always used to provide a cross-reference. Many current receivers use five satellite channels simultaneously, four to provide an accurate fix and the fifth to provide quick access to the correct codes as a new satellite appears above the horizon. The mobile antenna needs to have a hemi-spherical response to signals with an elevation as low as 5° above the horizon.

To descramble/despread the received signal the receiver has to generate a copy of the satellite code and multiply this with the incoming code. When the code patterns agree, the correlation detector ensures that the energy spread throughout the 2 MHz bandwidth becomes concentrated into the original 100 Hz baseband, with the significant coding gain. The recovered data is then automatically entered into the mobile terminal computer. This then solves the necessary four simultaneous equations in an iterative manner, before presenting the results to the user on a suitable display.

The Receiver: Because of the very low level of input signal, the use of a double-superhet receiver is practically mandatory. Conversion to a 1st IF in the range 100–200 MHz gives good image frequency rejection whilst high gain can readily be established in a 2nd IF of 5–20 MHz. Since all the satellite frequencies are multiples of the basic chip rate of 1.023 MHz, it is useful to use the same concept in the receiver. Thus a carrier frequency of

1540×1.023 MHz = 1575.42 MHz, and 1st IF of
120×1.023 MHz = 122.76 MHz or 160×1.023 MHz = 163.68 MHz

requires a local oscillator frequency (LO) of 1452.66 MHz or 1411.74 MHz.

By using such related frequencies, the locally generated signals can be synchronously locked. This avoids spurious beats and harmonics that could cause interference problems, leading to a reduction in processing accuracy. Fig. 6.31 shows how the decoding/despreading action is carried out in two stages, one section generating the correct GOLD code tracking whilst the other synchronises the locally generated carrier to the incoming signal. This, due to the relative velocities of the transmitters and receivers is affected by Doppler shift modulation. Processing at this stage is therefore usually accomplished by using either a phase lock or Costa's loop circuit.

The code tracking loop of Fig. 6.32 shows basically how the correct decoding sequence is selected by comparing the recent past and the predicted future codes. This results in the 2 MHz wide IF signal being reduced to its original 100 Hz baseband, but still in the BPSK format. After demodulation this can be processed to yield the navigational information.

Differential GPS: This technique greatly enhances the accuracy of fixes without affecting the integrity of the military part of the system. It requires the additional use of a fixed GPS receiver located to within a surveyed positional accuracy of better than 1 metre. The computer in this reference receiver compares its accurately known position with that obtained from GPS so that it can calculate a correction factor. This is then transmitted to all suitably equipped mobile GPS receivers, where it can be applied to the raw GPS fix to calculate a position, accurate to within 5 metres. This extra feature requires that the GPS receiver must be equipped to receive the correction factor transmissions either via satellite or over a conventional radio link.

GLONASS (GLObal NAvigation Satellite System) (USSR)

This system operates in a very similar manner to NAVSTAR, in that it employs spread spectrum ranging techniques, but uses multiple L(S) band carriers in the range 1602.5625 MHz to 1615.5 MHz. The system will eventually employ 24 operational satellites in three inclined orbits. Again, two levels of location accuracy can be achieved using a C/A code of 511 kbits and a P code of 5.11 Mbits. The code repetition rate is also 1000 times per second but with a basic bit rate of 50 kbit/s.

Developing Systems

Because all the systems described operate in very similar frequency ranges and have many features in common, the use of an associated computer

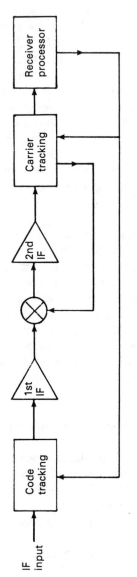

Fig. 6.31 NAVSTAR signal processing stages.

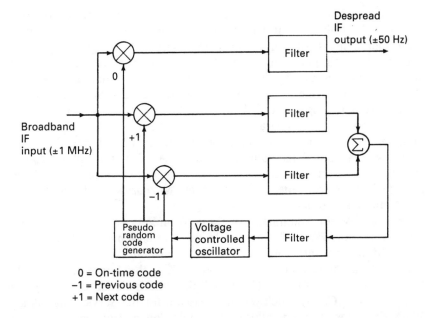

Fig. 6.32 Code tracking system.

allows many of the services to be integrated into one common receiver. In particular, the introduction of the low-cost, high-processing speed transputer, allows many of the hardware problems of this integration to be solved using computer software. Receivers capable of operating from both NAVSTAR and GLONASS are possible; INMARSAT-C terminals can be integrated with both; and hand-held portable receivers have been demonstrated to be a practical proposition.

Radio Determination Satellite Service (RDSS) (32)

This commercial system is used to report accurate information about the location of mobile units back to transport operator's headquarters. It is suitable for use on road vehicles, railway trains, aircraft or boats and can provide positional fixes accurate to less than 10 metres. In addition, provision is also made to allow for short, two-way messages to be passed between headquarters and the mobile user. The system operates through either low earth orbiters (LEO) or geostationary satellites, using either the standard LORAN-C (LOng Range Aid to Navigation), TRANSIT or GPS positioning services.

In 1986 the US GEOSTAR Corp was granted a licence and the first system became operational in 1987. In the same year, a European consortium LOCSTAR signed a licensing agreement as the first step to a global expansion of the service. Since that time many more countries including Australia, Brazil, China and India have adopted the concept.

Development is still progressing in stages. The original System 1 operated through spare transponders of the ARGOS system LEOs known also as NOAA 9 and 10. This provided basic positional reporting and allowed for only a one-way data service. System 2 went through a number of enhancement stages during 1988 and 1989, providing a more accurate location service plus two-way messaging. System 3, which provides for satellite ranging and navigation, came into service during 1991 and provides a positional fix accurate to less than 7 metres, plus two-way messaging. System 4 is due to come into operation during 1993/94 and provides for a world-wide network that allows *roaming* and *visiting*, with many millions of users.

The 1987 World Administrative Radio Conference for Mobile Services (known as MobWARC'87) agreed to the international standards jointly proposed by the International Telecommunications Union (ITU), the Federal Communications Commission (FCC), the International Radio Consultative Committee (CCIR) and the International Civil Aviation Organisation (ICAO). Such was the demand for this service that almost 100 MHz of spectrum was allocated in the following frequency bands:

> 1610 – 1626.5 MHz, L Band user up link
> 2483.5 – 2500 MHz, S Band user down link
> 5150 – 5216   MHz, C Band satellite to ground station link
> 6525 – 6541.5 MHz, C Band ground station to satellite link

The system uses two radio channels, one for data and messages from the mobile to the central control station (in-bound) and the other in the reverse direction (out-bound). Each message sent over the system carries the unique identity code (ID) of the user, thus ensuring that only the addressee receives that message. Because RDSS uses spread spectrum, burst transmission techniques, many users can communicate simultaneously with the geostationary satellite. The GOLD spreading codes form the basic protection from data collision in this CDMA controlled system. The GEOSTAR control centre transmits interrogation signals through the satellite to all users many times per second. When received by a mobile, it replies with its own unique ID together with the digital data that it has waiting. The mobile transmitter then automatically shuts down whilst the control centre stores the messages in its computer. Once stored, the control centre transmits an acknowledge signal and then forwards the message to a prearranged destination.

The basic packet structure consists of about 130 000 time slots or chips. Each packet has a maximum length of 128 bytes of which 105 are allocated to messages. A packet consists of header, address, data or text, CRC error check, and trailer, with small variations between in-bound and out-bound messages. Messages that cannot be contained within a single packet require the addition of a sequence number.

The user terminal automatically counts from the start of a frame and, if prompted, it transmits a prearranged chip sequence. This prevents other users from accessing the same slot. Generally three-chip separation is allowed between all users existing in the same frame. If any user terminal is blocked in any given frame it has a better than 99% chance of access within the next frame. A tier of priority is provided in the protocol, so that a terminal with an emergency can obtain rapid access.

Binary PSK is used for the primary coding format at a basic rate of 15.6 kbit/s. The use of a 1/2 rate convolution code with constraint length 7 doubles this to a transmission rate of 31.2 kbit/s. Burst lengths vary between 20 ms and 80 ms, and the use of a GOLD spreading code expands the bandwidth to 16.5 MHz. The typical transmit power level is in the order of 50 watts (17 dBW).

As an example of the value of RDSS to transportation generally, a fully operational System 3 installation can handle 4 million out-bound, and more than 20 million in-bound messages per hour. For added security and privacy, the system also provides for message encryption.

A typical user terminal consists of a main receiver/processor/transmitter, a keyboard/display unit, receive/transmit antennas, plus the option of a hard copy printer. During the few years of system development, the size and power requirement of the terminal has reduced markedly and hand-held units are now a possibility.

The antennas used vary from a 42 cm monopole for LORAN reception to microwave patch and helical devices for L and S band.

6.6 SATELLITE NEWS GATHERING (SNG)

Basically there are two situations where the use of mobile TV camera equipments are needed. One is the large international event, well publicised beforehand, which can be covered readily with relatively large earth stations with 3 or 4 metre diameter antennas and using standard FM TV satellite transmissions. Alternatively, emergency-type news situations may require the rapid movement of equipment and operating staff over very great distances. Satellite equipment to support such an operation requires a much higher degree of portability. The equipment should be capable of being packed into flight cases for transit with the operating crew using scheduled airlines.

Both C and Ku Bands are available to provide such a service and, of the two, Ku Band is probably the most attractive as there are fewer terrestrial microwave telephone networks to produce mutual interference problems. In any case, signals transmitted over Ku Band can easily be retransmitted or networked over C Band, where the need arises.

For operation in remote areas, primary power consumption can be a critical factor. The antenna size/gain and high power ampilifier (HPA) power requirements form one compromise. The antenna size, beamwidth and side lobe response also has to be taken into consideration. With satellite spacings of 2° or 3°, it is important that the side lobe response should be well within the CCIR 29–25 logθdB. Portability limits the HPA perhaps to 300 watts maximum. A typical 600 watt HPA weighs something like 150 kg and needs more than 5 kw of primary power.

The use of digital processing with some form of bandwidth compression can also be considered. This, combined with forward error correction, could mean that the transmit EIRP required to produce a satisfactory link performance might well be reduced. If the bandwidth is reduced, then so is the noise power contained within it. Thus the signal power can be reduced to maintain the same S/N ratio for a given channel capacity (Shannon's theorem). Reducing the EIRP requirement automatically improves the portability.

A digitally encoded broadcast standard colour TV signal would require a bit rate of around 180 or 220 Mbit/s for NTSC or PAL respectively. A bit rate reduction could impair the displayed picture quality, unless use is made of the high level of redundancy contained in the signal. The average information in the TV picture varies relatively little from frame to frame, except in the areas containing movement. This suggests that a compromise might be made between link margin, data rate and picture quality.

It has been conjectured that, in terms of link margin, the standard FM satellite link of 30 MHz is roughly equivalent to a bit rate of 60 Mb/s. Thus a reduction to 8 Mb/s or even 2 Mb/s would yield an improvement in link margin of 60/8 to 60/2, or about 9 to 15 dB. With operation towards the edge of the satellite coverage area and in bad weather, the S/N ratio might well be degraded. However, using FEC techniques the digital signal would be less affected, any deterioration in the display quality being largely restricted to moving edge definition.

Two equipments that meet these operational and portability criteria are the Ranger (18), manufactured by Alcatel Multipoint Ltd, Witham, Essex, UK, and the Mantis (19) manufactured by Advent Communications Ltd, Chesham, Bucks, UK. Both are available in various configurations to meet a range of different conditions. Depending upon configuration, they are designed to be packed into four to seven flight cases for transportation purposes, and deployed and operated by a two-person crew.

Ranger is equipped either with a 1.5 metre or 2.4 metre diamond-shaped antenna, as shown in Fig. 6.33, whilst Mantis employs either a 1.5 metre or 1.9 metre diameter circular antenna (see Fig. 6.34). The Ranger antenna is of the off-set fed type, whilst Mantis uses a prime focus feed. Both

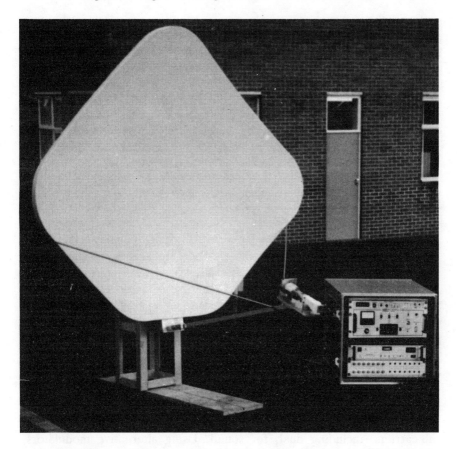

Fig. 6.33 Ranger satellite news gathering equipment (courtesy of Alcatel Multi-point Ltd, Witham, Essex).

antennas are sectionalised for ease of transportation, with the panels being accurately aligned and securely clamped when deployed. Both adequately meet the side lobe response requirement of all satellite operators.

Both systems provide a minimum EIRP of 69 dBW with the smaller antenna and a single 300 watt HPA. The addition of a phase combined second HPA raises the power output by almost 3 dB. This, together with the larger antenna, then raises the EIRP to maximum of 75 dBW.

The equipments (see Fig. 6.35) include a receiver used for both antenna alignment and link monitoring. Both transmit and receive signals are processed in a double conversion mode, with transmission in the range of 14 to 14.5 GHz. The receive band is split into two sections: 10.95 to 11.75 GHz and 12.25 to 12.75 GHz. Transmission uses standard frequency modulation, plus energy dispersal for the vision channel, and a choice of

Fig. 6.34 Mantis satellite news gathering equipment (courtesy of Advent Communications Ltd).

sub-carriers, including dual, for sound. Using alternative modulators, provision is made to handle digital signals from 64 Kbit/s to 2 Mbit/s, plus those from a video digital coder/decoder (codec).

Provision is made to transmit either NTSC, PAL or SECAM standard TV signals, using either frequency modulation or the enhanced video compression codec at bit rates of 1.5 to 8 Mbit/s.

Referring to Fig. 6.35, the signal input to the digital video encoder may be either composite video or RGB, plus audio, from either an electronic newsgathering (ENG) camera or a video tape recorder (VTR). These signals are converted into a common digital data stream with a very low bit rate, from 1.5 Mb/s for the North American and 2 Mb/s for the European standards, up to 8 Mb/s. The video sources may be either PAL or NTSC encoded. In fact, the codec is capable of working between the 625/50 and 525/60 line standards without using the conventional analogue video standards convertor. The encoding process involves a technique known as 'conditional replenishment', which consists essentially of transmitting only the frame-to-frame changes in the picture information. Picture areas containing movement are predictively encoded using DPCM, followed by a

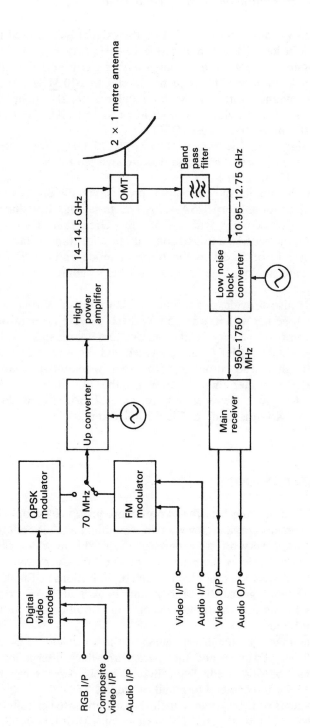

Fig. 6.35 Block diagram of typical SNG system.

statistically matched variable length code. This data signal is then used to quadrature phase shift key (QPSK) modulate a 70 MHz carrier.

Alternatively, when the network demands a higher picture quality, the analogue video signal can be used to frequency modulate a 70 MHz carrier in the conventional manner. Either 70 MHz modulated signal is then up-converted and amplified to provide an output at 14 to 14.5 GHz from the antenna via the ortho-mode transducer (OMT).

The down-link signals in the range 10.95 to 12.75 GHz, which are handled in two sub-bands, are filtered and down-converted by the LNB to provide a first 1F of 950 to 1750 MHz.

An alternative to the mobile SNG concept is provided by the S23 system of Midwest Communications Corp USA. This system is housed within a 30 m^3 road transport truck body, with a 2.6 metre Gregorian offset fed antenna mounted on the roof, in a folded configuration for transportation. The antenna has a gain of 50 dB in the transmit mode and 49 dB for receive, with adequate sidelobe rejection to meet the standards set by all satellite operators.

This size of body also provides for a limited facility outside broadcast (OB) service. This, together with its own 15 KW diesel powered generator, allows the system to provide a very flexible service to network stations, in any TV format. Two 300 watt HPAs are provided and these can function either in parallel for increased output or with one as a redundant spare. The maximum EIRP is better than 77 dBW for the 14.0 to 14.5 GHz transmission band (Ku), with the receiver covering 11.7 to 12.2 GHz. For the audio service, provision is made for dual sub-carriers.

6.7 VIDEOCONFERENCING

A videoconference is a *meeting* between two or more groups of people in different locations, which is organised via a two-way sound and vision link. The locations may be in very different parts of the world, in which case satellites can play a vital role in the provision of the link. Currently, the space links for these services are provided by such organisations as Eutelsat and Intelsat, with the large earth stations and back-haul links being provided by the PTTs and other national telecommunication organisations over wide band land lines.

The major claims made for the effectiveness of these services lie in the reduction of travel for personnel and the speed with which international videoconferences can be organised. Typically, up to six persons can be comfortably catered for at each end terminal or studio.

The essential elements of this system, which is based on closed circuit TV (CCTV), are shown in Fig. 6.36. (20). The system consists of two *face-to-*

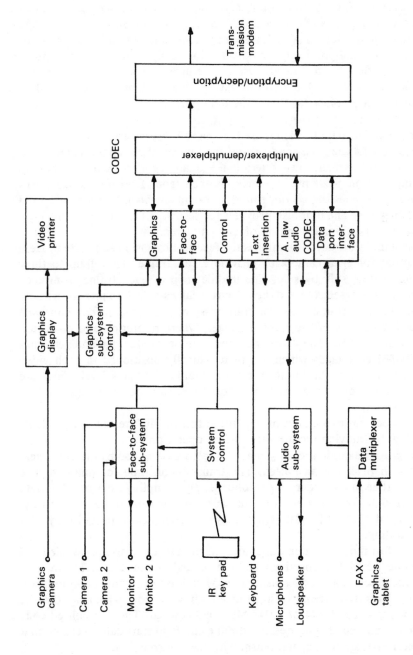

Fig. 6.36 Block diagram of video-conferencing terminal.

face cameras and two colour TV receivers for the incoming signals. A local colour monitor is also provided and the terminal is often housed in a roll-away cabinet. The TV system for each end terminal as described here, can be either PAL or NTSC encoded. (Provision is also made for SECAM coding to be included.) A high resolution *graphics* camera is also incorporated. Additional provision can be made for an alphanumeric keyboard to insert text data as the need arises, together with a video printer and interface for a VTR. A digital interface is also available to connect a Fax machine or a graphics tablet via the data port.

Local control over the system is via a conventional infra-red remote control, similar to that which is used on many TV receivers. This can also be used to control the *focus* and *zoom* functions on the graphics camera.

The central element in each terminal is again a codec, made by Marconi Communication Systems Ltd under licence from British Telecom. This codes and decodes the vision, sound and data signals that are transmitted over land lines. The unit also provides for security of information by using optional encryption/decryption of the data stream.

The main information channel is the face-to-face vision sub-system. Here it is usual for each camera to provide a full frame view of up to three persons, sitting side by side. The middle strip of each frame from each camera output is selected and then electronically stacked, to provide a three-over-three, full frame view for transmission. This *split screen* approach avoids the corners of the camera images, where resolution and distortion are not of the best. The decoder receives this stacked image and generates two half-sized images which are then centrally positioned in each video frame, with the top and bottom sections blanked to black level. These are then displayed on two adjacent monitors to give a wide screen image of the distant studio.

The codec has two modes of operation: fast for high resolution graphics, and slow for moving picture face-to-face operation.

The graphics mode uses a systematic replenishment coding system, where every 19th pixel is transmitted. The transmission time for one complete frame is then in the order of 2 seconds. Initial sampling rates of 12.5 MHz and 4.166 MHz are used for luminance and chrominance respectively. Adequate definition for detailed display of a document is provided by 6 bit PCM coding.

In the fast mode, the luminance and chrominance signals are sampled at 5 MHz and 2.5 MHz respectively and then coded in 8 bit PCM. The moving areas are detected and coded using DPCM and this generates a non-uniform bit rate. This data stream is thus buffered in a memory which can then be read out at either 1.544 or 2.048 Mb/s as required. The transmitted code combines moving area addresses with line and field start data to ensure that the receiver maintains synchronism with the transmission.

The audio channel, which can provide for stereo as well as mono signals, uses A-law companding and PCM coding, for *sound-in-vision* (SIV). This is a time multiplex used to avoid the necessity for separate sound and control channels. The audio signal is sampled at a rate synchronised to the line frequency of the local camera, and the derived data words are inserted, one in each line blanking period. The sound sub-system also includes an *echo cancelling* unit. Echoes can be caused by the outputs from the terminal loudspeakers which may be retransmitted by the terminal's microphones. This unit samples the incoming audio signal, compares it with the studio sound and estimates the echo component, which is then subtracted from the outgoing audio signal.

6.8 BANDWIDTH COMPRESSION VIDEO CODEC

This codec, for which two typical applications were described in sections 6.6 and 6.7, has to respond to two basic situations: the transmission of both moving pictures of reasonable quality and high quality images of documents (21,22,23,24). As the codec operates on the components of the video signal, i.e. luminance and two-colour difference signals (Y and U,V in the PAL system), it is possible to cater for the various TV system formats such as NTSC, SECAM and PAL. Because these formats chiefly only differ in the way that the colour difference signals are processed and transmitted, the variants can be handled by the use of suitable video interfaces. An RGB interface is also provided, but the data compression only operates on the video components (Y and U,V). Provision is also made for the processing of both audio and data signals.

A block diagram of the encoder section of this codec is given in Fig. 6.37. Its companion decoder is essentially complimentary.

Sound in Vision

To improve flexibility of operation, the codec may be installed at a PTT's switching centre. In this case, the video signal will be delivered over a land line. In order to avoid the use of separate sound and data channels, a time multiplex, known alternatively as sound in vision (SIV) or sound in syncs (SIS), can be used. The land lines used to carry the video signal would be those normally used to carry thirty-two 64 Kb/s (2.048 Mb/s) PCM channels in Europe, or twenty four 64 Kb/s channels plus 8 framing bits (1.544 Mb/s) in the USA. For this application of SIV, the audio signal is low-pass filtered, and sampled at 7.8125 KHz (1/2 line rate) and quantised using 8 bits. These are then inserted into a $2\mu S$ time slot during each alternate line sync pulse period. The instantaneous bit rate is thus 4 Mb/s,

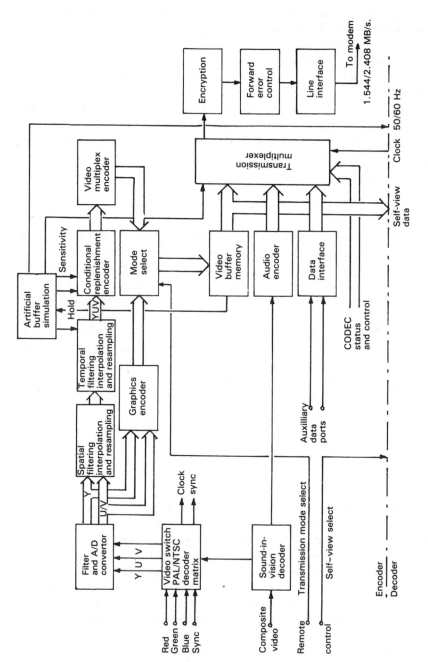

Fig. 6.37 System block diagram for encoder section of video codec.

with an average rate of $(8 \times 7.8125 \text{ KHz})/2 = 31.35 \text{ Kb/s}$. On the alternate lines, 4 bits can be transmitted for identification purposes, leaving 4 bits spare for future expansion.

Redundancy in the video image

Bit rate reduction techniques make use of the high level of redundancy contained in the information of a television picture. The literature (25,26,27) explains very clearly how this is related to the picture definition and quality. Even in its analogue form, the picture has been derived by a sampling process. Horizontal scanning represents sampling in the vertical sense, as this divides the image into very narrow strips, the smallest definable strip being represented by the number of lines per picture height. A series of separate images or frames can represent a moving scene, the movement between images going undetected. The frame rate therefore represents a sampling of images per unit time. Digitisation of each scan line would introduce horizontal sampling.

The definition or degree of fineness of detail contained in an image is measured in terms of pixels and the rate at which these can change in intensity. This therefore leads to a response that can be evaluated using three kinds of frequency:

- Horizontal frequency f_x, as cycles per picture width, c/pw
- Vertical frequency f_y, as cycles per picture height, c/ph
- Temporal frequency f_t, indicating how fast the image intensity is changing as a function of time.

These components can be represented in three-dimensional space by three frequencies mutually at right angles to each other. Obviously, the greater the number of pixels used and change per unit time, the greater the bandwidth needed for transmission. For bandwidth compression to be effective this detail must be minimised in some way, such as by combining pixels. This can be achieved if each frame is digitised and the image data stored in a semiconductor memory.

In both the analogue and digital sense, filtering amounts to the combination or averaging of information, in this case from a number of different pixels. Horizontal or vertical filtering is achieved by combining pixels from the same or adjacent lines respectively. Temporal filtering is achieved by combining only those pixels that occupy the same position in successive frames. By including more than one kind of pixel in the same filtering process, horizontal-temporal or vertical-temporal filtering can be achieved. The general term for this is *spatio-temporal* filtering. From this, it can be seen that there are two distinct areas where redundancy can occur: from line to line or *intra-frame*, or from frame to frame or *inter-frame*. Both can be invoked to reduce the bandwidth required, by transmitting the stationary area detail less often and/or by signalling the position of the moving areas.

Operation of Codec

Irrespective of the format in which the video signal is received, it is ultimately coded in the Y, U, V components of the PAL format. For the moving image mode, these components are filtered, sampled and quantised into 8 bit PCM, as indicated by Fig. 6.37. The sampling frequencies are 5 MHz and 2.5 MHz respectively for the Y and U,V components. The colour components are processed in a separate channel similar to that shown for the luminance in Fig. 6.37. In the high resolution graphics mode, the corresponding sampling frequencies are 12.5 MHz and 4.167 MHz, but the coding used is 6 bit PCM, plus 2 bits *dither*, to minimise *edge contouring* which is likely to occur under low-level and noisy signal conditions. Since only every 19th pixel data is transmitted, a complete image takes about 2 seconds to build up. Therefore any movement that is introduced will appear fuzzy.

The core of the bit rate reduction process is contained in the conditional replenishment encoder shown in Fig. 6.38. A spatio-temporal pre-filter processes the image data and noise, to improve the performance of the movement detector. Moving areas are predictively coded in DPCM and a statistically matched address code of variable length is added to identify the particular areas. Frame stores are required at each end of the link and the algorithm used depends upon these two stores remaining in track. As the technique used generates a non-uniform data rate, regulatory action is taken to smooth out the data rate to match the channel capacity. The buffer memory is used to smooth out short-term peaks in the data rate, but the coding algorithm can also be adaptive when needed.

If the encoder generates data faster than the channel rate, the buffer will fill and might overflow, with consequent loss of information. Feedback paths are included within the codec to control this situation. Control is exerted by either:

(1) Reducing the sensitivity of the movement detector circuit as the buffer fills.
(2) Using temporal sub-sampling, whereby alternate field data is discarded, leaving the decoder to interpolate between adjacent fields.
(3) Using spatial or pixel sub-sampling where certain moving area data is discarded, again leaving the distant decoder to interpolate, but between adjacent pixels.
(4) If these actions fail to prevent overflow, then all moving area data can be suspended with a *freeze frame* display until the buffer starts to empty.

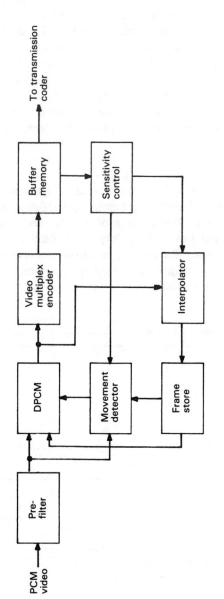

Fig. 6.38 Block diagram of conditional replenishment encoder.

Synchronism is assured because both line and field start codes are included in the signal multiplex.

The transmission multiplexer of Fig. 6.37 organises the digital data components into frames, which are in turn assembled into 16-frame multiframes and 8-multiframe super-multiframes for both European and USA transmission standards. A European frame structure is shown in Fig. 6.39. Although the USA frame structure is similar, it only contains twenty-four 64 Kb/s time slots. Whereas the European frame is synchronised from time slot zero, the USA frame is synchronised by the 193rd bit in each time slot. For cross system working (NTSC to PAL), the European time slots 26 to 31 are vacated to allow a remultiplexer to transform the frame formats.

Primary Code Format

The primary coding for the transmitted bit stream is alternate mark inversion (AMI) in the USA and high-density bipolar 3 (HDB3) in Europe. In both cases, a *zero* is coded by the absence of a pulse and a *one* by alternate positive and negative pulses of 50% duty cycle. Such a format is most suitable for cable and other forms of transmission, because their frequency spectra contain no dc component. However, a long string of zeros can lead to loss of clock synchronisation and consequent data errors.

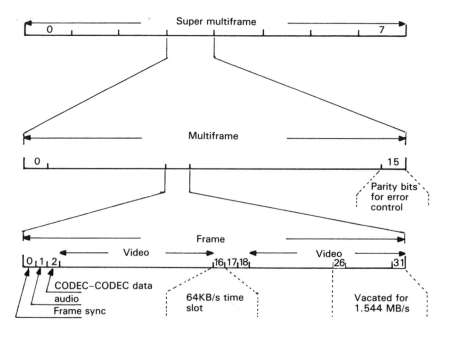

Fig. 6.39 Transmission multiplex structure.

This problem can be minimised by the use of the bipolar N-zero substitution (BNZS) format, typical codes being B6ZS or B8ZS, where a string of six or eight zeros is substituted by bipolar pulses that infringe the 'alternate ones' rule. The decoder recognises these substitutions as code violations and produces the necessary corrective action.

The HDB3 code is similar to the B3ZS, where code violations are added for strings of four consecutive zeros. The HC-5560 Transcoder chip by Harris Semiconductors Ltd is an example of the way in which integration has developed. This IC can work with either AMI, HDB3, B6ZS or B8ZS, simply by changing a 2 bit code on its control lines.

Coding and Decoding Algorithm

Since the encoder transforms data from a non-uniform generation rate into a constant transmission rate, the receiving decoder has to perform a complimentary transformation. The received data is stored in a memory, and the actual data held is used to control the rate at which it is to be read out to reconstruct the video image. This is partly achieved via the synchronising line and field start codes and by the periodical transmission of the level state of the encoder buffer, ensuring that the decoder buffer readout follows at the correct rate. Figure 6.40 shows a cluster of pixels from successive video lines and frames. Even in a moving area, small movements will mean that significant correlation exists in both the spatial and the temporal senses. In general, the difference between the data for pixels X and X", will be similar to the difference between other

(a)

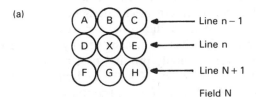

(b)

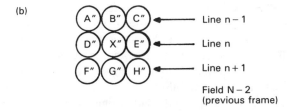

Fig. 6.40 Cluster of pixels on successive frames.

corresponding and surrounding pixels. This feature is used for both the predictive coding before transmission and interpolation at the receiver. An algorithm that is based on this premise, and gives good video images is;

> If $\mid$ D-D$''\mid$ > T$_h$ then $\mid$ (C + D)/2 $\mid$ can be used for X, the data for C and D being already available.
>
> If $\mid$ D–D$''$ $\mid$ < T$_h$, then X$''$ from the previous frame can be used for X, T$_h$ being a predetermined threshold value.

Optional Facilities

Although many satellite links provide encryption for data security, it may be necessary to add such protection for the land line section. This is provided for in the codec.

Forward error correction in Europe uses a BCH code (4095,4035), whose 60 parity bits provide for the correction of up to five random errors plus one burst of up twenty-six errors in each block of data. The parity bits are inserted into the last eight time slots of each multiframe. For the USA, a shorter code is used, the parity bits for which are transmitted in frame time slot 2.

H.261 Codec (see also Section 5.11, Transform Coding and 7.5, Digital HDTV Systems) (24 and 35)

International collaboration through the Motion Picture Experts Group (MPEG) and Joint Photographic Experts Group (JPEG) committees of the CCITT has led to the development and design of a codec that is ISDN compatible. It can be used for audio visual services at bit rates between 64 Kbit/s and 2 Mbit/s. This has been achieved by allowing each codec (transmitter and receiver) to apply pre and post processing of the local television signal. Either 625/25 or 525/30 standard formats generate a *common image format* (CIF) signal for the transmission channel. The CIF is based on 288 non-interlaced lines (sequential or progressive scanning) per picture and 30 pictures per second. 625 line systems use 576 active lines per picture (frame), so that 288 lines form one field. Therefore 625/25 codecs only have to perform a conversion to meet the 30 Hz picture rate and 525/30 codecs already operate at the correct frame rate so these only have to convert between 240 and 288 active lines. The advantage of the CIF is that all transmissions are to a common standard, a feature that means that the core codec design is to a world standard and this considerably aids interconnectability.

The CIF is formed by sampling the video signal luminance component at 6.75 MHz (chrominance at 3.375 MHz) to produce 352, 8 bit samples per line. This is equivalent to an analogue bandwidth of about 3.4 MHz, so that the processed image is only about 30% below that of full studio quality.

A lower image resolution is acceptable for applications such as video phones. For these cases a second standard is incorporated in H.261. This is based on 176 samples per line and 144 lines per picture. Since this represents half-resolution in each dimension the format is referred to as *quarter common image format* (QCIF).

The coding scheme and data compression algorithm is based on *differential PCM* (DPCM) and *discrete cosine transform* (DCT) with *motion compensation*. Conceptually, the 1D (one-dimensional) DCT is equivalent to taking the Fourier transform and retaining only the *real* or cosine part. The 2D (two-dimensional) DCT can be obtained by performing a 1D operation on the columns of a matrix, followed by a second 1D operation on the rows. The sampled YUV (luminance and chrominance) components of the video signal are processed separately but in the same way. The basic processing technique for each component is shown in Fig. 6.41 which indicates how the previously coded image data is subtracted from the current one to produce a *difference* image (point 1). The image is divided

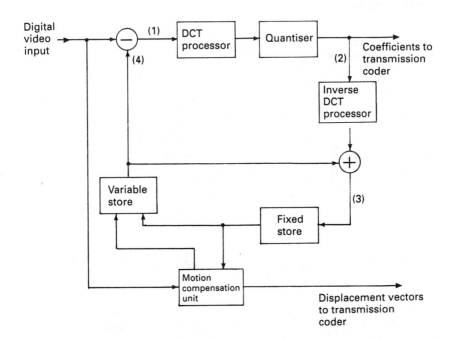

Fig. 6.41 Basic elements of H.261 codec encoding.

into blocks of 8 × 8 pixels which are then DCT processed using the transfer function:

$$F(u,v) = \frac{1}{4} \sum_{i=0}^{7} \sum_{j=0}^{7} f(i,j) \, CosA \, CosB$$

where $A = \dfrac{(2i + 1)u\pi}{16}$ and $B = \dfrac{(2j + 1)v\pi}{16}$

i,j & u,v = 0, 1, 2, 3 . . . . 7 and
i,j are the spatial coordinates in the image plane,
u,v are the corresponding coordinates in the transform plane.

The coefficients in each block represent the amplitudes of the various spatial frequencies present, the top left-hand element in this matrix, representing the average or dc level for the whole block. The matrix values are then scanned in a zigzag fashion as indicated in Fig. 6.42 to produce a serial bit stream. At this stage only the dc component and a few low spatial frequencies have any significant magnitude. The following quantising stage sets all the very low values to zero and truncates the others to a set of preferred levels, further reducing the number of values that need to be transmitted.

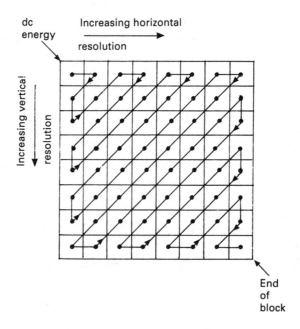

Fig 6.42 Zigzag scan of 8 × 8 pixel block.

Motion compensation is achieved by minimising the frame to frame differences that have to be transmitted. This operates by taking four blocks (16 × 16 pixels) from the current image that has been reconstructed via the inverse DCT processor, and searching over ± 15 pixels vertically and horizontally to find the best match. The variable length store is then adjusted so that the best match section of the image is used for subtraction at point 4. The vertical and horizontal translations derived from this *block matching* are later included in the transmission multiplex as motion correction or displacement vectors. Whilst the data bits are acquired at a variable rate and need to be transmitted at a constant rate, buffer stores are needed at both encoding and decoding stages. Control signals are therefore added, to ensure that the receiver decoder uses up the data bits at the correct variable rate.

At the receiver, the decoder uses the DCT coefficient values and the displacement vectors with a *block matching* technique, to reconstruct the image in a manner similar to that at point 3 in Fig. 6.41.

Data compression for transmission is achieved by using Run-length coding (RLC) that functions in the following manner. The quantised ac coefficients usually contain many runs of consecutive zeros and these are indicated by two 4 bit nibbles as follows. The upper 4 bits indicate the number of consecutive zeros before the next coefficient, whilst the lower 4 bits indicate the number of significant bits in the next coefficient. Following this code symbol are the significant bits of the coefficient, whose length can be determined from the lower 4 bits of the code symbol. At the receiver, the inverse RLC decoder translates the coded bit stream into an output array of ac coefficients. It takes the current code and inserts the number of zeros indicated by the RLC. The coefficient entered into the array has the number of bits indicated by the lower 4 bits on the RLC and a value determined from the number of the trailing bits.

For still pictures a further degree of bit rate reduction can be achieved by using a modified Huffman code. In this case, shorter code words are used for frequently occurring symbols and longer codes for the occasional ones. This can be based on a look-up-table stored in a read-only-memory (ROM) or alternatively operated in an adaptive manner for specific images. This involves creating a table based on the frequency count of each symbol in the image. This data table is then stored in a random-access-memory (RAM) for use in a similar way.

Although the digital processing in the video channel is fast, the propagation delay has to be matched in the audio channel to maintain lip-sync.

6.9   SMALL SYSTEM SATELLITE SERVICES (S⁴)

The technical developments of the last decade have had a marked effect on both satellite-borne equipment and ground station characteristics. Improvements in semiconductor technology now provide amplifier devices that operate up to higher frequencies, with higher gain and lower noise factors. Improvements to high power amplifiers (HPA) for ground stations and the travelling wave tube (TWT) and solid state RF power amplifiers for satellite transmissions, contribute to the need for smaller antennas. Due to the availability of suitable integrated circuits (ICs) the efficient modulation schemes and coding techniques of a decade ago have now become a practical reality. The traditional system design trade-offs using Shannon, Hartley and Nyquist criteria still apply. The above improvements give an adequate S/N ratio (particularly with digital processing) to allow the use of low power transmissions using small antennas. The new systems are capable of providing significantly high data rates with acceptably low bit error rates. As an example of the extent of development, hand-held radio-paging devices have now been demonstrated to be practical.

Micro-satellites

These small satellites are the result of cooperation between Amateur Radio Satellite Corp (USA), the University of Surrey (UK) and Surrey Satellite Technology Ltd (UK), and form part of the PACSAT Communications Experiment (PCE). They occupy a low earth polar orbit with a period of 90 to 100 minutes, so that any earth station is covered at least three times every 12 hours. The satellite carries a full scale *store and forward* messaging system, with messages and data being stored in an on-board 4 Mbyte RAM disk. The current satellite carries an on-board *charge-coupled-device* (CCD) image sensor, but other devices may be used. The stored messages and data are transmitted using a packet technique for distribution to the many scattered outposts, the system software being IBM-PC compatible. Due to their relatively very low cost micro-satellites form a useful concept for testing new ideas under actual space conditions.

Briefcase SNG Terminal

This small, light, go-anywhere system was developed by the cooperation between Columbia Broadcasting System (CBS), Teleglobe Canada and Skywave Electronics Ltd. Weighing only 14.5 kg, it is capable of providing good quality speech and still pictures plus data into any PSTN (public switched telephone network) over a satellite channel. It will operate in the full duplex mode over INMARSAT-C links of 1.5 to 1.6 GHz to a hub earth station which itself operates in the 4 and 6 GHz band.

The antennas consist of two flat patch microwave arrays, providing circular polarisation with a gain of 17 dBi. Together with a 10 watt RF power amplifier stage, the system produces an EIRP of 27 dBW in a 5 KHz bandwidth. This allows two terminals to operate simultaneously over a single 25 KHz INMARSAT channel. The antennas are fixed to the lid top which is orientated in azimuth and elevation using locking hinges. During deployment the antenna is aligned using the satellite beacon signal. Speech transmissions employ *amplitude companded single sideband* (ACSSB) and the digital modulation scheme uses DMSK at 2400 bit/s.

Power for the terminal may be provided by locally rechargeable NiCd batteries or by 110/230 volt 50/60 Hz mains supplies. It is anticipated that operation over a higher power satellite would allow an even smaller terminal to be produced.

When deployed, the terminal can also receive calls, but with the distant operator dialling first the hub terminal code, followed by the briefcase terminal code.

Developments in Personal Communications Networks (PCN)

The following two examples of *blue sky* research indicate the way in which satellites are likely to affect the expansion of personal communications capabilities. Both are largely due to the fact that such services can reach target areas, that are impossible or uneconomic by means of terrestrial services.

Iridium system (element 77 in the atomic table)
This proposal, originally made by Motorola Corp, is being considered by a consortium that additionally includes British Aerospace, Lockheed Aircraft Corp and INMARSAT, as a possible global network that could be in operation as early as 1996. The system will be based on 77 low earth polar orbiters at an altitude of about 660 km. The satellites will be positioned within 7 orbits, 11 to an orbit, and each equipped with inter-satellite communications links.

Calls will be made from handsets similar to the current cellular telephones, switched between satellites, then ultimately to a ground station to be patched into the terrestrial telephone system.

It is anticipated that frequency spectrum in the range of 1 to 2 GHz, will be allocated to this service during the next International Telecommunications Union (ITU) conference in 1992.

For low population densities and remote rural areas, this service which is intended to be compatible with, and complimentary to, INMARSAT should prove to be more economic than extending the present terrestrial telephone system.

The Japanese Experiment

During 1992, Japan plans to launch the ETS-V1 satellite into geostationary orbit. This will carry five experimental systems based on millimetre wave transmissions. Frequencies in the range 35 to 45 GHz have been chosen, largely due to the high gain available from small antennas. The experiments are:

(1) Slow speed SNG working at 512 kbit/s, with the ground terminal operating from a 40 cm diameter antenna and uplink power of 2 watts.
(2) Portable geophysical and meteorological data gathering and monitoring stations for natural phenomena.
(3) Video telephony/conferencing.
(4) Personal communications services based on 64 kbit/s, for a wide range of applications around the Pacific area.
(5) A mobile radio service for travellers in road vehicles.

The up-links will operate in the range 43.5 to 47 GHz, whilst unusually, the down-links will occupy the lower range of 37.5 to 39.5 GHz. Mostly the ground stations will use 30 cm diameter antennas, whilst those on the space craft will be 40 cm diameter. With gains in the order of 37 dB and 39 dB respectively and 0.5 watts of RF power, the EIRP will be in the order of 37 dBW. This should produce a received power level at both satellite and ground terminal of about −143 dBW. All experiments will operate in the single channel per carrier (SCPC) mode.

Very Small Aperture Terminals (VSAT)

VSAT systems are usually considered to be those designed to operate with antennas that are the equivalent of 1.8 metres diameter or less. The relatively low gain provided by these has to be countered by very careful design. Reference to the rules of communications (Shannon, Hartley, Nyquist), will show that systems operated in this way suffer from a reduced signal to noise (S/N) ratio and hence an increase in error rate. However, an acceptable error rate can be achieved by reducing the information transmission rate. VSAT systems are therefore essentially designed to provide digital services at a reduced data rate, typically 9.6 kbit/s, a value that can rise to 1.544 or 2.048 Mbit/s under special circumstances. Second generation VSAT terminals operate as transceivers, working through a satellite and a higher powered ground or hub station equipped with an antenna of 3.5 metres diamater or more. The larger antenna and greater power level is necessary to maintain an adequate overall S/N ratio. The hub station provides the link to other network users, either via a second satellite link, or the terrestrial PSTN or other suitable network. Due to improving technology, third generation VSAT's are able to operate directly in the VSAT to VSAT mode, without the intervention of a hub station.

VSAT networks form a valuable global link or bridge between isolated terrestrial local area networks (LAN) with which they have much in common and can easily be integrated. They thus add considerably to the communications capability of international corporate users, even to the extent of providing private video/telephony conferencing. For example, the K-Mart Corp in the USA operates a large in-store network serving some 2100 stores. This is used for the fast validation of credit cards, inventory and stock control and many other functions that need to be addressed by such a large organisation. The Chrysler Corp operates a system with similar features but adds the facilities of including video training programmes for staff in remote areas, plus the rapid availability of technical information for the service staff. In Europe, the Daimler Benz organisation has established a network that is ISDN compatible and extends into eastern Europe. The system allows for 64 kbit/s channels to provide a full range of services, even to the extent of renting spare capacity to outsiders.

VSAT systems operate both in C and Ku band, but the latter probably carries the greater part of these services. The systems may use any of the techniques of modulation, access and control that apply to satellites and LANs.

Terminals are divided into two basic sections as indicated by Fig. 6.43. The outdoor section comprising low noise block convertor (LNC) and high power amplifier (HPA) are usually contained within a single casing and mounted in the manner shown in Fig. 6.44. The HPA which is a semiconductor device, commonly produces an RF power of some 2 to 3

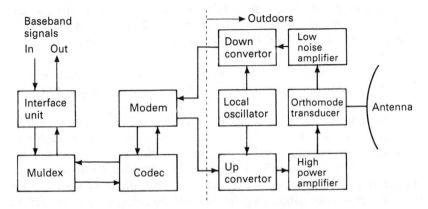

Fig. 6.43 Typical VSAT earth station.

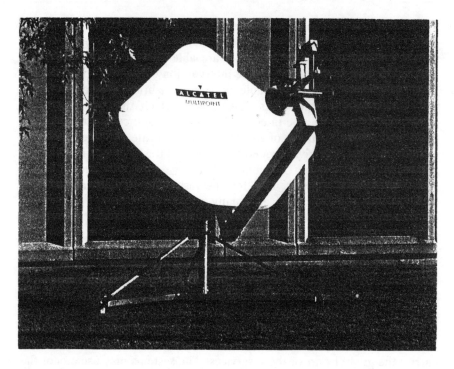

Fig. 6.44 Out-door unit VSAT terminal (courtesy of ALCATEL Multipoint Ltd, Witham, Essex).

watts. This assembly also carries the orthomode transducer and waveguide filters. These are necessary to separate the transmit and receive signals. Both transmitter and receiver function on the dual conversion, double superhet principle. The local oscillator is normally a crystal controlled phase lock loop, with frequency synthesis to select the correct frequencies for up and down link operation. The typical 1st IFs are around 1 GHz for Ku band and 600 MHz for C band, whilst the 2nd IF is commonly either 70 or 140 MHz.

The typical figure of merit (G/T) for the front end is in the order of 18 to 20 dB/K, depending on the combination of antenna size and low noise amplifier characteristics.

The indoor section will contain the second stage of frequency conversion, plus the modem, codec and muldex (multiplexer/demultiplexer), these elements providing complimentary functions in the transmit and receive modes. The interface unit processes the baseband signals and matches the various input devices to the digital processing stages. The muldex combines/distributes a given transmission channel between several inputs. The codec and modem stages handle the forward error correction (convolution code), and convert the source code, usually pure binary, into the channel code, either BPSK or QPSK, the latter being preferred because it doubles the channel capacity without increasing the bandwidth requirement.

Fig. 6.44 shows the outdoor components of the Alcatel Multipoint Ltd VSAT terminal equipped with a 1.5 metre diamond-shaped reflector antenna. (A 2.4 metre antenna is optionally available.) The degree of portability achieved by such systems is clearly shown. This system is designed to operate with the Alcatel FASTAR network which includes a hub station equipped with an antenna of at least 5 metres diameter.

The basic network provides 20×9.6 kbit/s SCPC trunks from VSAT to hub and a single TDM frame of 20×9.6 kbit/s timeslots in the reverse direction. The VSAT to hub channels consist of 1 random access signalling channel and 19 SCPC DAMA (demand assigned multiple access) message channels. The reverse direction timeslots are similarly allocated, 1 for signalling and 19 DAMA for messages.

The system control operates on a modified SPADE protocol. (See System Access Protocols.) A VSAT accesses the hub via its signalling channel f_1 requesting connection. The hub informs the calling station of the assigned channel via its signalling timeslot t_1 of the TDM frame. The VSAT then transmits on frequency f_x and receives on timeslot t_x. For VSAT to VSAT links, 2 DAMA channels need to be allocated by the hub, f_x and t_x as described above, and f_y and t_y for the relay between the hub and the called VSAT.

Using the system software, FASTAR can be programmed to provide the higher data rates of 19.2 kbit/s or 64 kbit/s, but with fewer DAMA channels.

System design

The link power budget can be calculated in the manner shown in Table 6.1. The number of VSATs that can be supported by a pool of DAMA channels is determined in the same way as the number of telephones that can be supported by the lines of a telephone exchange. The ERLANG* loading of each VSAT is calculated and summed to give the network loading. The network size is adjusted for the grade of service required in terms of probability of queueing for a channel. Typically 20 DAMA channels will support 500 to 1000 VSATs with 2% queueing probability. Thus some VSATs can support a single terminal whilst others support a network of telephones and computers. A VSAT supporting a local network may need simultaneous access to more than one DAMA channel.

---

*ERLANG, the unit of traffic intensity. One permanently engaged circuit has a traffic flow of 1 Erlang. If the average number of simultaneous calls in progress in a given period over a given network is E, the traffic intensity on that network is E Erlangs. (Erlang-hour: one Erlang of traffic intensity maintained for one hour.)

TSAT (33)

A third generation VSAT system, so called because it has been designed to operate at the CCITT T1/E1 data rates of 1.544 Mbit/s (North America) or 2.048 Mbit/s (Europe). Unlike the VSAT that normally operates in a *star* mode of connection, the TSAT is capable of operating in a *mesh* configuration, connecting each location to all others without the need of a hub station. Not only are these systems faster with a higher data capacity, they are also significantly cheaper. The networks, originally conceived by SPAR Aerospace of Canada, are ISDN compatible in that they provide both basic rate access of 2B + D (64 kbit/s) and primary rate access of 23B + D (1.544 Mbit/s) or 30B + D (2.048 Mbit/s) (B channel used for voice/data services and D channel for signalling purposes).

System Access Protocols

Generally, the traffic loading on a VSAT terminal tends to be very variable, so that the systems tend to be bandwidth rather than power limited. The access problems are very similar to those of LANs and therefore the choice of protocol, which can be critical, needs to be carefully chosen bearing the following points in mind:

- Average acceptable transmission delay
- Range of data rates involved
- Acceptable message failure rate
- Message format; fixed or variable length.

SPADE (Single-channel Per-carrier DEmand assigned and multiple access)
A control station monitors the channel frequencies in use, then answers a request for interconnection by allocating a free channel to the two stations.
ALOHA (System originally devised by University of Hawaii for interworking of computer networks)
With the pure ALOHA protocol, terminals transmit as and when they have data ready for transmission and also continually monitor the network. If two terminals transmit together, they detect a collision of data and both back off for a random time period before trying again. This is basically a collision detect-multiple access (CDMA) system.
Reservation TDMA: Each terminal is allocated a particular time slot of the time frame for its transmissions.
Slotted ALOHA: As in pure ALOHA but terminal transmissions are restricted to allocated time slots.
Selective spectrum ALOHA: An ALOHA protocol reserved for use with spread spectrum transmissions, when packets of data will not be affected by collisions.
ACK/NACK protocol (acknowledged/not acknowledged): When the receiving station decodes a packet without error it transmits an acknowledgement signal. If the transmitting station does not receive this acknowledgement, it

decides that an error has occurred and automatically retransmits the data packet.

Reservation TDMA is particularly useful when relatively long transmissions are required. Slotted ALOHA has an advantage over pure ALOHA in that once a terminal has accessed a time slot, it has sole use of this until the end of transmission. Selective spectrum ALOHA has the advantage that data collisions are not problematic but the system has to include some form of forward error correction (FEC). ACK/NACK system is not very useful over satellite links. Due to the overall time delay of about 250 ms, the following block may be transmitted before the lack of acknowledgement has been recognised and this will produce a data collision.

6.10 GLOBAL BUSINESS SERVICES

In the main, the major owners of the satellite space segments are organisations such as INTELSAT (International Telecommunications Satellite Organisation), EUTELSAT (European Telecommunications Satellite service, an arm of the European Space Agency) and PASAT (Pan American Satellite Inc). These then lease the day-to-day operation to other suitably qualified organisations, who are often but not always the national PTTs. These operate ground stations with antennas that may range in diameter from 3.5 to 13 metres. The transponder channels may then be sub-leased to other end users. The services provided include, video, voice and data and these are often linked into the national PSTN or ISDN. In fact, apart from the propogation delay, the high grade of service makes the satellite channel practically transparent to the end users.

Large Aperture Satellite Systems (LASS)

The wide range of antenna sizes in operation allows for a correspondingly wide range of LNAs/LNBs to be used to meet the systems overall G/T ratios. The parameters and standards for the complete ground stations, including uplink, downlink and footprints (specified down to a minimum elevation angle of 10°) are closely specified in the literature (34). Power budget calculations and a comparison of the relative merits of TDMA versus SCPC are explained in Table 6.1.

Services are provided in both C and Ku bands and the latter is probably most popular outside of the USA. The transponder bandwidths are in the order of 36 to 72 MHz and these may be occupied by either analogue or digital signals. All forms of analogue modulation are catered for typically using FDMA/SCPC. Most digital services carried use QPSK/2-4PSK modulation with TDMA/SCPC protocols. The bandwidth for these is usually allocated in contiguous blocks of 64 kbit/s depending upon the user requirement up to a maximum of 8.488 Mbit/s. This, however, can be extended up to 24.576 Mbit/s for special cases using differential encoding and demand assignment (DA). FEC using 1/2 or 3/4 rate convolutional coding plus Viterbi decoding is commonly employed to produce a bit error rate better than 10^{-3}. A form of spread spectrum transmission is commonly employed using scrambling to act as a form of energy dispersal to obtain full channel occupancy. Even so, it is still allowed for the originator to include encryption within his own domain.

Specialised Satellite Service Operators (SSSO)

In the UK a number of operators have been licensed to operate fixed and mobile ground stations through most satellites. Of these organisations, Satellite Information Services (SISLink) was probably one of the first operators to lease transponders on a permanent basis. SISLink provides satellite news-gathering services and news feeds on an international basis; a private racing network of video and sound distribution to bookmakers throughout Europe; and videoconferencing and other corporate communication services. Although the racing network uses a B-MAC system developed by Scientific Atlanta Inc for vision with Dolby ADM for sound, all other forms of modulation and television formats can be handled.

REFERENCES

(1) Bhargava, Haccoun, Matyas and Nuspl 1981 *Digital Communications by Satellite (Modulation, Multiple Access and Coding)*. New York: John Wiley and Son Inc.

(2) European Space Agency, ECS Data Book esa BR-08.

(3) European Space Agency, ECS Data Book esa BR-08 Appendix.

(4) Knowles, K.A. Muirhead Office Systems Ltd, UK. Private communications to author.

(5) Knowles, K.A. Muirhead Data Communications Ltd UK. Private communications to author.

(6) Turner, J. *The Down Link Signal Characteristics of TIROS-N Satellites*. National Meteorological Office UK.

(7) Turner, J. *Orbital Parameters and Transmission Frequencies of Meteorological Satellites*. Oct. 1985. National Meteorological Office, UK.

(8) European Space Agency (1981) *Introduction to the Meteosat System, SP-1041*.

(9) European Space Agency (1980) *Meteosat WEFAX Transmissions*.

(10a) European Space Agency (1984) *Meteosat High Resolution Image Dissemination*.

(10b) Baylis, P.E., Mather, J.R. and Brush, R.J. (1980) *A Meteosat Primary Data User Station*. Research carried out for ESA by the University of Dundee, Scotland.

(11) Kirsch, R. (1986) *Maplin Magazine*.

(12) Rowland Smith, P. *WSR 513 Service Manuals*, Vol. 1 and 2. Feedback Instruments Ltd, (UK).

(13) Kennedy, D.M., International Maritime Satellite Organization, (IMARSAT) London. Private communications.

(14) Marconi International Marine Co Ltd (1985) *'Oceanray' Satellite Communications Terminal Ship's Manual.*

(15) Appleyard, S.F. (1980) *Marine Electronic Navigation*, London: Routledge and Kegan Paul Ltd.

(16) Johns, A.G. *SAT-NAV 412 Operating Manual.* Birmingham: Walker Marine Instruments Ltd.

(17) Warman, M.R. and York, E.L. Polytechnic Electronics plc, Daventry, UK. Satellite Navigator, Advanced Service Information. Private communication to author.

(18) Player, J.K. and Baker, D.W., Multipoint Communications Ltd, Essex, UK. Private communication to author.

(19) Turrall, P.A.T., Marconi Communications Systems Ltd, Essex, UK. Private communication to author.

(20) Steele, G., SVT Video Systems Ltd, Essex, UK. Private communication to author.

(21) Carr, M.D., Clapp, C.S.K. *et al.* (1982) 'Practical problems of implementing a conditional replenishment video codec over an error prone channel'. *International Conference on Digital Image Processing.* York.

(22) Carr, M.D. and Clapp. C.S.K. (1984) 'The integration of television standards conversion into a conditional replenishment codec for visual teleconferencing'. *International Teleconference Symposium.* London.

(23) Nicol, R.C. and Duffy, T.S. (1983) 'A codec system for world-wide videoconferencing'. *Professional Video* (now *Broadcast Systems Engineering).*

(24) Kenyon, N.D. (1985) *British Telecom Technology Journal*, Vol. 2, No. 5.

(25) Annegarn, M.J.J.C., Arragon, J.P. *et al.* (1987) 'HD-MAC: a step forward in the evolution of television technology'. *Philips Technical Review* **43**, No. 8.

(26) Barratt, L.H. and Lucas, K. (1979) 'An introduction to sub-nyquist sampling'. *I.B.A. Technical Review* **12**. London: Independent Broadcasting Authority.

(27) Lever, I.R. Analogue to digital conversion. *I.B.A. Technical Review* 12. London: Independent Broadcasting Authority.

(28) McConnel, K.R., Bodson, D., Schaphorst, R. (1989) *FAX: Digital Facsimile, Technology and Applications.* London: Artech House.

(29) *Earth Observation Quarterly.* Journal of ESA, ESTEC. Noordwijk, Netherlands.

(30) Mattos, Philip G. 'Global positioning by satellite'. *Electronics + Wireless World*, Feb. 1989, p. 137.

(31) Spilker, J.J. 'GPS Signal Structure and Characteristics'. *Journal of the Institute of Navigation (USA)*, Vol. 25, No. 2, Summer 1978.

(32) Pierce, J. and Finley, M. Eds. (1989) *Understanding Radio Determination Satellite Service.* Geostar Corp, USA.

(33) Garland, Peter J., SPAR Aerospace Ltd, Canada. Private communication to author.

(34) Ackroyd, B. (1990) *World Satellite Communications and Earth Station Design.* London: BSP Professional Books Ltd.

(35) Clark, R.J. (1985) *Transform Coding of Images.* London: Academic Press Ltd.

Chapter 7

Television Systems

7.1 REVIEW OF CURRENT SYSTEMS

It is intended in this section only to review the principles and characteristics of the current system in order to understand why and how certain developments are occurring. For a complete understanding of three systems, NTSC (National Television Standards Committee), SECAM (Séquential á Mémoire), and PAL (Phase Alternation Line-by-line), the reader is referred to some of the standard works of reference (1, 2, 3, 11).

All the variants of the colour TV systems in service are based on the concept of the brightness or luminance (Y) component signal of the earlier monochrome systems, plus chrominance (colour) information. In all three systems, the Y component is formed from the weighted addition of red (R), green (G) and blue (B) gamma corrected camera voltages that represent the three primary colour signals, i.e.:

$$Y' = 0.299R' + 0.587G' + 0.114B'.$$

Three colour difference signals are needed at the receiver, $(R'-Y')$, $(G'-Y')$ and $(B'-Y')$. When the Y' component is added to each, a colour signal voltage is regenerated that represents the original colours.

Of the three colour difference signals, $(G'-Y')$ is always of the lowest amplitude and will thus be affected the most by noise in the transmission channel. Since the Y' component contains a portion of all three colours, $(G'-Y')$ can be regenerated at the receiver from the Y', $(R'-Y')$ and $(B'-Y')$ components. These three components are therefore all that is necessary for the transmission of colour TV signals. Thus the two colour difference signal components to be transmitted are of the form $(B'-Y')$ and $(R'-Y')$. These are modulated on to sub-carrier frequencies and in order to conserve bandwidth, the spectral components of modulation are interleaved with the luminance spectrum. In a limited way, this interleaving influences the choice of actual sub-carrier frequency. The major differences between the systems lie in the way in which the chrominance information is modulated on to the sub-carriers.

Historically the NTSC system of the USA was the first system to enter operational service. This was made compatible with the North American monochrome service that was based on a 525 line image format. The line and field rates were nominally 15.75 kHz and 60 Hz respectively. To accommodate the 3.579545 MHz sub-carrier and minimise the interference beats between this and the sound carrier, these frequencies were marginally offset to 15.734254 kHz and 59.94 Hz respectively. (These changes were insufficient to cause a monochrome receiver to lose timebase lock.) The colour sub-carrier therefore lies exactly half-way between the 227th and 228th harmonic of the line timebase frequency. The chosen colour difference components amplitude modulate quadrature (QAM) versions of the same carrier frequency, to produce I (in phase) and Q (quadrature) components, with the double side band, suppressed carrier (DSBSC) mode being employed. Like the Y' component, both I and Q signals are formed by weighted addition of the R', G' and B' primary components.

$$I = 0.596R' - 0.275G' - 0.322B' \text{ and}$$
$$Q = 0.211R' - 0.523G' - 0.312B'$$

As indicated in Fig. 7.1, both the I and Q components contain a fraction of both $(R'-Y')$ and $(B'-Y')$. This was arranged to combat the effects of non-linearity, where amplitude distortion might cause a change of colour saturation. More importantly, however, phase distortion would produce an actual change of hue (colour).

Consistent with the technology then available, the NTSC system provided very adequate colour images and maintained compatibility with the previous monochrome service.

The later developing European colour TV industry recognised these possible problems and devised the SECAM system in France and the PAL system in West Germany, almost in parallel. Due to technological advances made during the intervening time, both of these systems provided better quality pictures and so became competing European Standards.

The SECAM system adopted the technique of transmitting the two colour difference signals sequentially, on alternate lines, using frequency modulation of two different sub-carrier frequencies, thus ensuring that there would be no cross-talk between these two components. However, the two frequencies needed to be very carefully chosen to minimise the effect on monochrome areas of the image because unlike both NTSC and PAL, both of which use DSBSC, the carrier amplitudes do not fall to zero when both the $(R'-Y')$ and $(B'-Y')$ components are absent, as in the grey picture areas.

The two colour difference signals of SECAM are designated as $D_R = -1.902(R'-Y')$ and $D_B = 1.505(B'-Y')$ which are pre-emphasised before being used to frequency modulate the two sub-carriers of 4.40626 MHz and 4.25 MHz respectively. The magnitude of the coefficients are chosen so that total deviation is restricted to 3.9 to 4.75 MHz for both sub-carriers. The negative coefficient of D_R results in the two difference signals

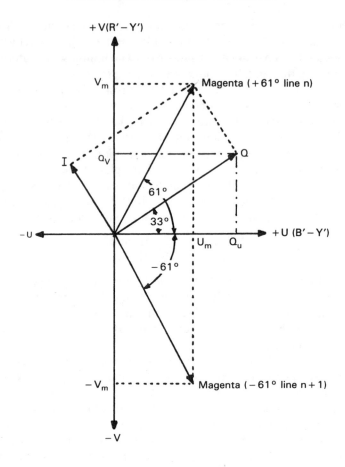

Fig. 7.1 Comparison of NTSC and PAL chrominance phasors.

deviating their carriers in opposite directions. This was adopted in order to minimise the possible effects of differential phase errors, but practice has shown this to be unnecessary. The later specifications for SECAM therefore use negative coefficients for both D_R and D_B. The sub-carriers are maintained to an accuracy within $\pm$ 2 kHz and represent the 282nd and 272nd harmonics of the line timebase frequency.

For the PAL system, the two colour difference components are scaled or weighted so that the total signal of luminance plus chrominance does not produce over-modulation. The two modulating colour difference components are thus

$$U = 0.493(B'-Y'), \text{ and } V = 0.877(R'-Y').$$

Like the NTSC system, PAL utilises QAM, DSBSC, but uses an alternating phase inversion, line-by-line, of the V (R' − Y') component. In this way phase errors on successive lines tend to average out and cancel, thus improving the colour performance. Fig. 7.1 shows how the NTSC and PAL chrominance components differ. Both the I and the Q signals contain an element of each colour difference signal due to the rotation of the reference phase. The figure compares the phasors for a magenta hue that forms the basis of the two chrominance signals.

The relationship between the sub-carrier and line timebase frequencies is much more complex due to the problem of *line crawl*. The precise value of 4.43361875 MHz represents 283.75 times line frequency plus an off-set of 25 Hz. This results in an eight field sequence before the colour sub-carrier phase repeats itself.

For all three systems, the sub-carrier frequencies have to be carefully chosen, not only for reasons of frequency interleaving but to avoid patterning due to beat frequencies between the luminance and chrominance signal components.

After some 30 years of development the increased size and brightness of the modern picture tube is causing other system impairments (*artifacts*) to become apparent, such as *cross colour/luminance* effects that arise due to the imperfect separation of the interleaved spectra. Luminance information can reach the chrominance channel and create false colours, and in a similar way chrominance information can create high frequency patterning in the receiver luminance channel. These effects can be minimised by the use of modern comb filter techniques or by digital processing of the composite video signal.

However, the sampled nature of the image signal creates aliasing problems that are not so easily solved. As was stated earlier the response of the image could be expressed in terms of three kinds of frequency, as indicated by Fig. 7.2(a). Considering the horizontal frequency f_x as a continuous function of time and hence analogue, the remaining two frequencies f_y and f_t represent the two-dimensional vertical and temporal sampling frequencies respectively. Fig. 7.2(b) represents this spatio-temporal spectra for a period when $f_x = 0$.

A sampling operation always generates repeated sideband pairs, related to the baseband and disposed around multiples of the sampling frequency. Fig. 7.2(b) translates such spectra into two dimensions simultaneously, where the *quincunx* areas (four corner and centre points of a square) represent the repeating spectra. The diagram is scaled for a 625 line system using 575 active lines per frame, giving a vertical resolution of 575/2 = 287.5 c/ph, with a 25 Hz temporal or frame frequency. Even if the spectral groups do not overlap to produce aliasing, the human eye will act an an imperfect filter. The approximate response of the eye is enclosed by

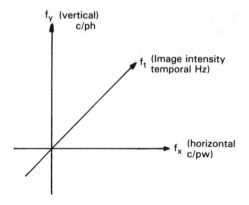

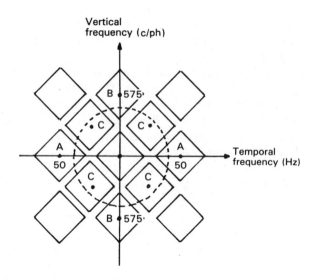

Fig. 7.2 (a) Three-dimensional representation of image response. **(b)** Two-dimensional spectra for $f_x = 0$.

the dotted line. This includes some of the repeat spectra, giving rise to an aliasing effect.

Referring to Fig. 7.2(b), area A is responsible for *large area flicker*, which is essentially a peripheral vision effect. It is related not only to the signal but also to the display viewing angle. Area B represents the *vertical aliasing*, which is responsible for the visibility of the line structure and the *Kell effect.*(4). This describes the loss of vertical resolution relative to that in the horizontal plane. Area C, the *temporal aliasing*, only exists in interlaced raster structures and is responsible for the apparent inter-line flicker or crawl. This effect also causes wheels to appear to rotate in reverse.

Increasing the frame rate will only increase the frequency at which this occurs.

Further technological improvements, in either of the current systems, are unlikely to remove these alias-based artifacts completely. So does the concept of a new, satellite delivered, TV service, offer the opportunity to introduce a *higher definition television* (HDTV) system? Several possible systems have been proposed and fewer have been practically demonstrated. The major debate for the last decade of this century will thus almost certainly revolve around the choices of either an *evolutionary* or *revolutionary* approach to devise an analogue or a digital system. HDTV is now defined as any system that is capable of subjectively providing a vertical resolution of more than 500 c/ph or 1000 lines.

7.2 IMPROVED PROCESSING OF CURRENT SYSTEMS

The areas where picture quality of the current systems can be improved are chiefly those areas that generate inter-modulation and other distortions through non-linearity, namely the IF amplifier, vision demodulator and colour separation/decoder stages. The use of integrated circuit versions of high gain IF amplifiers combined with surface acoustic wave (SAW) filters, followed by synchronous or phase-lock loop demodulators, is now common. There is thus very little scope for improvement in these areas. The luminance/chrominance separation can be improved by the use of charge coupled devices (CCDs) and gyrator circuits in IC form. (A gyrator is a two-port, non-reciprocal device whose input impedance is the reciprocal of its load impedance – for ·instance, a gyrator terminated with a capacitance, behaves as an inductance: $Z_{in} = R^2/Z_L$, where R is the gyration resistance.) Such an approach produces better results with NTSC than with PAL systems because of the simpler relationship between the sub-carrier frequency and the harmonics of the line scan frequency.

The comb filter, which is based on a delay line and adder, has been a feature ever since the inception of the PAL system. Difficulties of producing delays of the order of 64 μS has so far been resolved using ultrasonic glass delay lines, as these have been the only suitable devices. However, comb filters can also be constructed in digital form, a feature that becomes increasingly attractive as digital processing encroaches further into this analogue field. The comb filter effectively averages the information in at least two adjacent lines. The number of such lines increases with the order of the filter. The use of a filter with a too high order leads to loss of vertical resolution.

The digital filter which can be constructed from semiconductor amplifiers and delays can also be an averaging device, the output depending not only on the present input pulse but also on the previous ones. The major advantages of the digital filter are that they are particularly stable and can be reprogrammable.

The wide band luminance signal passes through its processing channel somewhat faster than the relatively narrow band chrominance signal. This leads to mis-registration of the two components. The inclusion of an additional luminance delay of around 700 nS provides some correction.

Scan Velocity Modulation (SVM)

This technique can be used with advantage on receivers with modern picture tubes. Figure 7.3(a) depicts the way that a part of the luminance signal is tapped off at its delay line, differentiated and then amplified before being applied to an auxiliary coil on the scanning assembly. Figure 7.3(b) shows how a luminance transient is enhanced by this technique. A luminance signal of the form $V_1 = A\mathrm{Sin}.\omega t$, when differentiated, becomes:

$$V_0 = \omega t A\mathrm{Cos}.\omega t$$

where $t = CR$, the time constant of the differentiator. This circuit will thus increase the slope of luminance transients to more clearly define changes of image intensity.

Colour Transient Improvement (CTI)

This technique uses the same principles as SVM but applied to the chrominance channel. Figure 7.4(a) shows the block diagram of an IC (Mullard TDA4560) specifically developed to provide this feature. Figure 7.4(b) shows the standard approach to matching the different time delays by the addition of a luminance delay line (i,ii,iii), whilst iv and v indicate how the enhancement of the chrominance component, by differentiation, improves the combined edge response.

A typical chip set that contains these features, designed for multi-standard receivers and produced by Mullard Ltd (UK), would include:

IF amplifier/demodulator	TDA 2549
Multi-standard colour decoder	TDA 4555/6
Colour transient improvement circuit	TDA 4560/1

Hybrid Processing

Other developments lead to hybrid processing – a combination of analogue and digital solutions – by using application specific ICs (ASICs) in certain suitable areas. Such an introduction can have many advantages, as it leads to the use of microprocessor control of the many new features available in

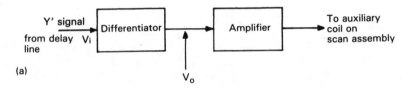

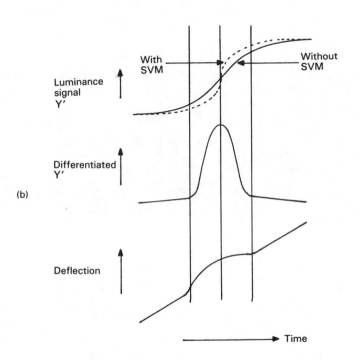

Fig. 7.3 Scan velocity modulation **(a)** principles **(b)** SVM waveforms.

modern receivers such as Teletext, Computer RGB Monitor, Freeze Frame, Video Printer, etc. The use of semiconductor memories as frame stores allows the same line or frame to be displayed twice in the same line or frame period, effectively doubling the scanning frequencies and removing flicker.

Taking this concept of the digital processing further, it becomes obvious that the earlier digitisation takes place, the lower will be the signal degradation due to the analogue circuitry. Conversion at RF is definitely not possible yet. Even at an IF of nearly 40 MHz, the improvements gained would not be cost effective. Therefore digitisation must follow immediately after vision demodulation.

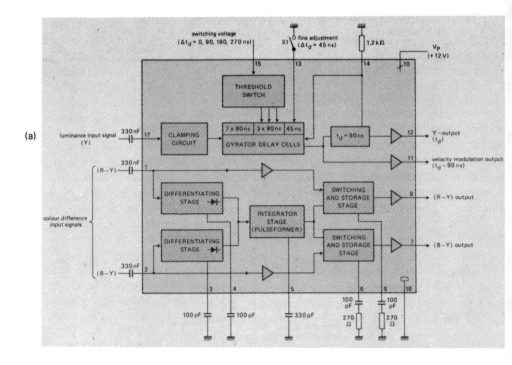

(a)

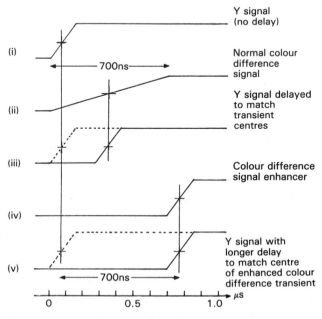

(b)

Fig. 7.4 Colour transient improvement.

7.3 DIGIT 2000 SYSTEM

The currently available receivers using the Digit 2000 system (5) represent the most advanced stage of the application of digital control and processing, available on the domestic entertainment market. User control is completely digital and the receivers provide for television with mono, stereo or bilingual sound channels. Video processing allows for the NTSC, SECAM and PAL systems of transmission, plus, the European Broadcasting Unions (EBU) new multiplexed analogue components (MAC) standards.

Referring to Fig. 7.5, the television signals are processed in the analogue format from the antenna through to the demodulator stages. Thereafter the processing is digital. This allows the use of advanced methods of video processing, which includes adaptive noise reduction and ghosting, aliasing and flicker suppression. Auxiliary processing allows for teletext and computer monitor applications, together with the additional inputs from video disc or cassette machines.

RF/IF Processing

Channel selection is controlled by frequency synthesis, with tuning steps of 62.5 KHz which are derived from division by 64 from a 4 MHz crystal-controlled oscillator. Automatic channel search is included, with AFC for fine tuning, this being carried out under software control from the microprocessor in the central control unit (CCU). The local oscillator frequency (f_o) is divided by 64 to produce a value within the operating frequency range of a PLL. This signal is compared with the required channel value, an action which causes up/down signals to be generated and fed to the tuner interface. These have a repetition frequency of 976.5625 Hz, of variable pulse duration and derived from 62.5 KHz/64. These two pulse streams produce a charge/discharge action on an integrating capacitor, which generates a tuning voltage proportional to the average of the two. The IF/demodulator circuitry uses *state of the art* integrated circuit technology for analogue processing. However, all of the tuning and alignment facilities normal in such circuits are carried out digitally. Initial alignment is carried out using a production line computer, to optimise the receiver performance. All of these tuning values are then digitally stored in an EEPROM. Subsequent servicing can involve the use of a service computer, to modify these values as necessary in order to re-establish the original performance. The demodulated output signals are multiplexed audio and composite video.

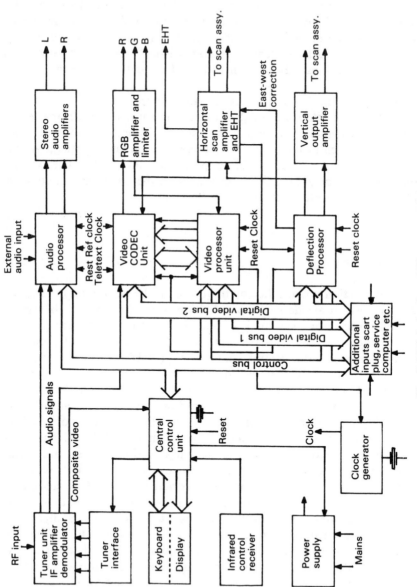

Fig. 7.5 Digital colour TV system (Digit 2000).

User Controls

Operation is controlled from either a front panel keyboard or an infra-red remote control sub-system. The latter uses a space duration 10 bit code, with start and stop bits to convey instructions. Logic 0 or 1 are determined by the time duration between very narrow pulses. Provision is thus made for $2^{10} = 1024$ different commands. These are usually split between $2^4 = 16$ differently addressed sub-systems, each capable of responding to $2^6 = 64$ different commands.

Audio Channels

Provision is made for either the West German $L + R/2R$ stereo mulitplex, or the USA's Zenith Corporation's multi-carrier $L + R/(SAP)FM$ multiplex. (SAP = secondary audio programme, carrying the L-R signal component.) Either multiplex is converted into a serial bit stream using *pulse density modulators* (PDM), the pulse density being proportional to the signal amplitude. The maximum pulse rate of 7.1 MHz means that no anti-aliasing filters need to be incorporated at the inputs. The serial data is converted into 16 bit parallel words for further processing. This includes, de-emphasis, tone, volume, loudness, stereo width and balance controls, plus the generation of pseudo-stereo from a mono signal. The final conversion to analogue format is carried out by two *pulse width modulators* (PWM). Such outputs only require simple filtering to provide the necessary drive to analogue power amplifiers.

Central Control Unit

This control unit is based on a microprocessor and is designed to work as an interface between the user and the receiver. The original factory alignment and tuning data is stored in the EEPROM and these values are used as references. Control is exercised over the three-wire serial control bus. (This includes, Ident, Clock and Data lines.) The programmability of the microprocessor allows different set manufacturers to design receivers around the chip set to meet their own particular specifications. Control signals to the various digital processors are passed over the serial bus, and these cause parallel data to be transferred as needed over one or other of the parallel data buses.

Video Codec and Processing Units

These two ICs work in conjunction to decode the composite video signal into its luminance (Y') and colour difference components (R'-Y') and (B'-Y'). The signal is first converted into a pseudo 8 bit parallel Gray Code (actually only 7 bits are generated). The Gray Code is particularly suitable at this stage. Each successive value changes only in one bit position and this feature assists the noise reduction which is carried out next. This signal is then passed over a parallel bus into the video processing stage, where it is converted into a simple 8 bit code for the luminance channel and an offset binary code for chrominance processing. In the luminance channel, the signal is filtered, the contrast level set and then delayed to obtain optimum luminance/chrominance registration for the display. The delay is achieved quite simply, by halting the movement of the luminance signal for a programmable number of clock pulses. The Y' signal is then passed back to the codec unit, still as 8 parallel bits. The chrominance signal is decoded into R'-Y' and B'-Y' components and also passed back to the codec unit, but in two 4 bit time-multiplexed groups. The three components of the video signal are then converted into analogue format and matrixed to produce the R,G,B primary colour signals.

In addition to inputs from external audio and video sources, interfacing is also provided for such auxiliary inputs as a SCART or Peritell socket and connection to the service computer.

Video Memory Controller

Figure 7.6 illustrates how the addition of a video memory controller (VMC) and six dynamic RAMs (DRAM) can be used to create a flicker-free picture. The 8 bit luminance and 4 + 4 bit time multiplexed colour difference signals from the VPU are input to the VMC. The 8 bit groups are reorganised into 24 bit parallel groups by an encoder, for storage in the six 64K × 4 bit DRAMs. This field data memory is organised into a 64K × 24 bit array. The read out process is practically complimentary, except that the field data is read out twice in each normal field period and then passed to the video codec unit for processing into RGB signals. The overall effect is to double both the horizontal and the vertical scanning rates to remove flicker. A freeze frame effect is simply achieved by halting the write to memory process.

Although this process yields good results with the normal video signals, a much faster rise time is needed in the analogue RGB stages when the receiver is used for teletext or as a home computer monitor. For these applications, an RGB double scan processor chip can be used to good effect.

Picture-in-Picture Processing (PIP)

The image data for the primary channel is processed as previously described (analogue video in Fig. 7.6). For PIP operation, an appropriate processor and two 16K × 4 bit DRAMs are added. The latter are used to store the

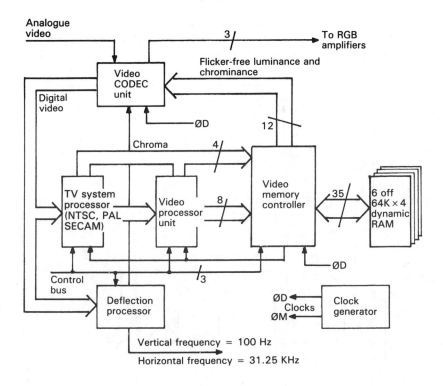

Fig. 7.6 Digit 2000 'flicker-free' image sub-system.

data for a secondary channel, whose image will be overlayed on that of the primary image. The data for this second channel is obtained via the receiver, video processor unit and TV system processor unit, the derived data being passed over data bus 2 (Fig. 7.7) to the PIP processing section. A sampling process selects every third pixel and every third line, to produce data for a 1/9th normal sized picture. This data is stored in the memory, to be read out during the appropriate part of each normal line and field scan period.

Closed Caption Transmissions

A simplified *teletext* type of service has been devised in cooperation with the American National Captioning Institute to provide a service for viewers with impaired hearing. Although this was specifically designed to operate with NTSC receivers, the service can be applied to any PAL or SECAM receiver that is equipped with a control microprocessor. It is thus

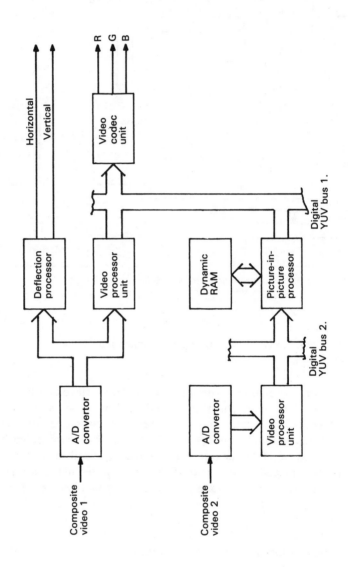

Fig. 7.7 Picture-in-picture processing.

compatible with the DIGIT 2000 system for which ITT have produced a single chip decoder, the CCD 3000.

An area of the screen is allocated for the display of up to 8 rows, each containing up to 32 characters. Apart from where characters are displayed, this area is transparent. For NTSC, the data is transmitted during line 21 of field 1 (odd) in NRZ format. The data stream at 503 kHZ, consists of a *clock-run-in* sequence followed by a *start code* consisting of two logic 0s and logic 1. This is followed by two 8 bit bytes that represent two ASCII characters. Simple parity check is provided by a single bit in each byte. The actual usable data rate is thus 16 bits per 1/30 sec or 480 bit/s. The received codes, including control codes, are decoded and stored in memory (RAM plus ROM character generator), to be displayed on the CRT under the control of the system microprocessor. A wide range of *fonts* and many of the normal teletext type of *attributes* is permitted.

7.4 MULTIPLEXED ANALOGUE COMPONENT SYSTEMS (MAC)

The opportunity provided by a new method of delivering television signals into the home is almost unique in the history of television engineering. It gives a chance to introduce a new TV system that overcomes the display problems of current receivers and to provide an image quality in keeping with the capability of the modern picture tube. A further influence on this opportunity for change presented by direct broadcasting by satellite (DBS) might be that provided by the introduction of *pay* or *subscription* TV. Whilst scrambling/encryption is possible with current services, it might be that this could be easier on a new system.

The use of frequency modulation for the space link has already been justified. Therefore a review of the effect of FM noise on the current methods needs to be made.

Figure 7.8 shows the disposition of the chrominance component within the luminance spectrum typical for current systems, the triangular spectrum for FM noise being superimposed. It can be seen that the chrominance component lies in the noisier region. Demodulation returns the chrominance signal to baseband together with the noise element. The colour response of the human eye is approximately triangular and complimentary to the FM noise spectrum. Therefore the colour noise appears in the most annoying region, being particularly troublesome in the highly saturated colour areas of the image.

Starting from this position in 1981, engineers from the UK Independent Broadcasting Authority (IBA) developed a hybrid solution, using a combination of analogue and digital processing for the vision and sound signals respectively. The Y and U,V components of video are time-compressed for separate transmission, thus automatically resolving the

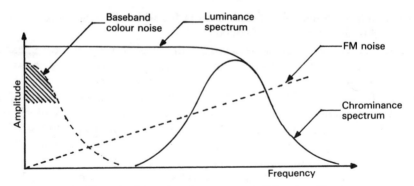

Fig. 7.8 Video signal spectra and FM noise effect.

cross luminance/chrominance problem. This concept, known as multiplexed analogue components (MAC), has given rise to a series of variants, each with some different properties. These have largely been developed within the framework of the European Broadcasting Union (EBU) and the Eureka-95 organisation of the EC (European Community). The format of the video line time multiplex, which is shown for two of the variants in Fig. 7.9, is typical of almost all versions. The active line period for the current systems is about 52 μS. For MAC systems, the luminance signal is time compressed by a ratio of 3:2 so that it occupies about 35 μS. The two colour difference signals are compressed by a ratio of 3:1, with a corresponding duration of approximately 17.5 μS. By transmitting the compressed Y signal on each line with one of the compressed U (C_B) and V (C_R) signals alternately on odd and even lines, the active line period is still 52 μS. As the action of the conventional line sync pulse can be replaced by a digital control, there is a period of about 10 μS which is available for digitised sound, data and digital control signalling.

A-MAC

This original version employed the analogue Y, U, V time multiplex, but with the sound modulated on to a separate carrier spaced 7.1 MHz from the vision carrier. Since this was likely to lead to intermodulation problems, A-MAC has not been adopted.

B-MAC

This version time-multiplexes the digital sound and data information at baseband into the 10 μs horizontal line blanking interval. The bandwidth is limited to about 6 MHz and so B-MAC is suitable for distribution over

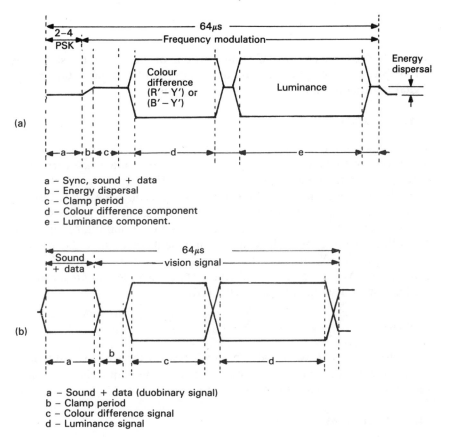

a – Sync, sound + data
b – Energy dispersal
c – Clamp period
d – Colour difference component
e – Luminance component.

a – Sound + data (duobinary signal)
b – Clamp period
c – Colour difference signal
d – Luminance signal

Fig. 7.9. (a) C-MAC (Packet) line multiplex; (b) D/D2-MAC (Packet) Systems line multiplex.

both satellite and cable links. The format has been adopted in Australia for use over the AUSSAT satellite, to provide television and data to the Homestead and Community Broadcasting Service (HACBSS). The format has also been extensively developed and marketed by Scientific-Atlanta Corp for use in the corporate communications business, even to the extent of providing a high-definition capability.

C-MAC

This uses an RF time division multiplex. The carrier is frequency modulated by the analogue vision signal components for 52 μS and then digitally modulated during a further 10 μS by a multiplex of the sound channel and data.

C-MAC/Packet

This is a derivative of C-MAC and forms one of the acceptable standards for DBS in Europe. The variation applies only to the organisation of the digital information into *packets* each of 751 bits.

D-MAC/Packet

This member of the family as originally specified, used an RF multiplex for the vision and sound plus data signals. The two components are carried on separate carriers, typically spaced by 10.5 MHz. These carriers could either be frequency modulated for transmission through a satellite link, or amplitude modulated (vestigial sideband) for a cable network. The current specification calls for the sound and data to be time multiplexed into the 10 μs line blanking interval, with all components being modulated on to the same carrier. Whilst the vision signal is frequency modulated in the conventional MAC manner, the digital components use *duo-binary* digital modulation at a sampling rate of 20.25 MHz.

D2-MAC

This variant of D-MAC has also been adopted by the EBU as a standard, chiefly for use in France and West Germany. The major change being a reduction in the sound plus data bit rate, by a factor of 2.

Studio-MAC (S-MAC)

This variant has been adopted in certain American TV studios that use component (Y,U,V) distribution within the production areas. Here the luminance is time-compressed by 13:6 and both colour difference signals by 13.3. This allows for the luminance and both chrominance components to be included in each line multiplex, conventional line and field sync pulses being retained. To ensure synchronism of the MAC clock circuit, a short burst at about 4.5 MHz (288 times line frequency) is included in the *back porch* period.

T-MAC

This is a French variation designed to be used for *electronic news gathering* (ENG) operations. The 3:2 time compressed luminance signal is transmitted on every line and the 3:1 compressed chrominance components C_B and C_R are transmitted line sequentially.

ACLE (Analogue Component Link Equipment)

This format was designed for use on current distribution links that are not capable of supporting the 12.5 MHz bandwidth requirement of S-MAC. For SNG (satellite news gathering) signals, the bandwidths of the luminance and chrominance components are typically restricted to about 4 MHz and 1 MHz respectively. This allows time compressions ratios of 3:2 and 6:1 to be used for luminance and chrominance.

D2-SMAC

This version was developed for the satellite transmission of television signals that will be used as a cable network feed. Such signals need to be received with a higher C/N ratio than would be expected for a direct DBS service. The component signals are produced in the same manner as D2-MAC, on which this format is based. The aim is to reduce the transmission bandwidth without loss of S/N ratio, so that four TV signals can be multiplexed into a single 36 MHz transponder bandwidth. This is achieved by sub-sampling (the S in SMAC) the luminance signal, discarding alternate samples and time compressing in the ratio of 2:1. This allows time to include in the multiplex, an *adaptive interpolation control signal* (AICS) that will be used to instruct the receiver how best to reconstruct the missing sub-samples. This is achieved in the receiver by interpolating either vertically or horizontally between adjacent pixels, according to the setting of an AICS bit. This is derived at the encoder by averaging the error of the four surrounding pixel interpolations and comparing with some predetermined threshold level. Depending upon whether the AICS bit is set to 0 or 1, the decoder will interpolate in the direction of minimum error.

Standard MAC Parameters

All the MAC versions eliminate cross colour/luminance, reduce colour noise, and give displays that are significantly better than anything achieved with either NTSC, SECAM or PAL. In addition, they provide for multiple sound channels, including multiple stereo, bilingual sound, teletext and controlled access TV.

For the European 625 line services, the following standard (6) has been adopted for C/D-MAC/Packet systems.

Frame frequency	25 Hz
Line frequency	15.625 KHz
Interlace	2:1

Aspect ratio	4:3
Y compression ratio	3:2
U,V compression ratio	3:1
Luminance bandwidth	5.6 MHz
Chrominance bandwidth	2.8 MHz
Transmission baseband	8.4 MHz (luminance 5.6 × 3/2 = 8.4)
	(chrominance 2.8 × 3 = 8.4)
Sound channel	40 to 15000 Hz
Sampling frequency	32 KHz
Dynamic range	>80 dB.

Whilst Fig. 7.9(a) shows the general arrangement of the time multiplex used for the MAC systems, the periods a, b and c, specifically refer to the C-MAC/Packet variant. Figure 7.9(b) shows the corresponding line multiplex for the D/D2-MAC/Packet variants. The digital components within period a include line and demodulator sync, sound and data, which are organised into packet multiplexes of 751 bits. Each packet contains a 'header' and 'data' section. Each header contains a 10 bit address to identify any one of 1024 different services, 2 continuity bits to link successive packets of the same service and 11 protection or parity bits for a Golay 23,12 cyclic error correcting code. This can correct any three errors in up to 23 bits. Packet 0 is permanently allocated to the service identification system.

The useful data area contains one 8 bit byte to indicate the packet type (PT). For instance, the sound decoder needs to know whether the coding law is linear or *near-instantaneously companded*, and also whether the sound signal is full bandwidth (high quality) or reduced bandwidth. The remaining 720 bits (90 bytes) may contain an *interpretation block* (BI) with the necessary instructions for setting up the sound decoder, or an actual sound data coding block (BC).

A triangular energy dispersal signal of 25 Hz at an amplitude suitable to provide a peak-to-peak deviation of 600 KHz is added. This is gated to zero during sound, sync and data periods. The clamp period allows the use of a receiver clamping circuit to remove this waveform at demodulation.

Frame Multiplex

Figure 7.10 shows the structure of the transmission multiplex for both C-MAC/Packet and D-MAC/Packet systems. Luminance information is carried on lines 24 to 310 and 336 to 623. Chrominance is additionally carried on lines 23 and 335, when the luminance period is set to black level. Line 624 can be used for carrier recovery and the setting of the receiver AGC system. Line 625 is completely reserved for frame sync and service identification. The field blanking period thus occupies 25 line periods.

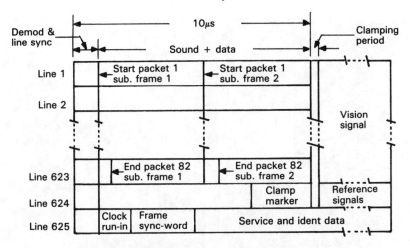

Fig. 7.10 C and D-MAC (Packet) systems frame multiplex.

Line 625 carries the maximum of 1296 bits, and with a duration of 64 μS this represents a bit rate of 20.25 Mb/s. On the remaining lines, the first 7 bits are used for demodulator and line sync whilst the remaining 198 bits are used for sound and data purposes. There are thus (623 × 198) + 8 = 123362 bits/frame which are organised into 164 non-contiguous packets, each of 751 bits, within the two sub-frames as shown in Fig. 7.10. Thus there is a significant number of bits that are unused and available for future developments.

In the D2-MAC/Packet variant, most of the 10 μS period is used to transmit 1 sub-frame of digital data instead of two, thus reducing the data bit rate from 20.25 Mb/s to 10.125 Mb/s.

Figure 7.11, which has been courteously provided by IBA UK, shows the C-MAC / Packet signal as displayed without expansion and remultiplexing on a standard receiver. The digital data, chrominance and luminance components can be clearly recognised. In particular, the chrominance section shows the sequential nature of the colour difference signals.

C-MAC/Packet Decoder

The RF/IF stages of the receiver provide FM processing appropriate to the satellite TV service. The video and audio/data signals are demodulated to baseband before being presented to the decoder that is notionally depicted in Fig. 7.12. Several of the functions, such as decoding of the service identification code and the data service (teletext), have been omitted for simplicity.

The baseband video signal is clamped to remove the energy dispersal component, and then de-emphasised. It is then sampled to separate out the luminance and chrominance components, but in digital format. Time

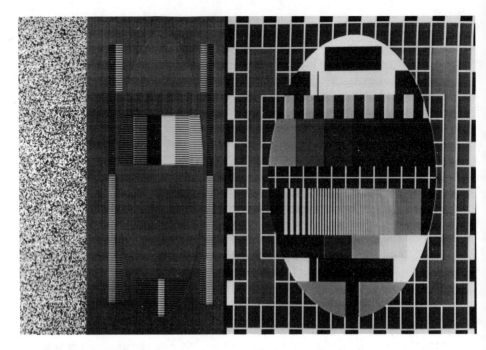

Fig. 7.11 C-MAC (Packet) signal displayed on PAL TV receiver (courtesy Independent Broadcasting Authority).

expansion is achieved by loading the two stores at the received data rate of 20.25 Mb/s and then reading out at the lower rates of:

$$20.25 \times 2/3 = 13.5 \text{ Mb/s for luminance and}$$
$$20.25 \times 1/3 = 6.75 \text{ Mb/s for chrominance.}$$

This process is controlled from the line and frame sync signals, obtained from the sound/data multiplex, in a manner that ensures the correct time registration of the Y and U,V components.

After low-pass filtering to remove the sampling frequency, the signals are matrixed to produce the RGB primary colour signals. Alternatively, the Y and U,V components can be recoded to either NTSC, SECAM or PAL standard, for display at a lower picture quality on a standard receiver.

The instantaneous data burst rate is 20.25 Mb/s. It has been shown that each frame holds 121680 bits, which are transmitted in 40 mS. The average bit rate is thus;

$$121680/40 \text{ mS} = 3.042 \text{ Mb/s.}$$

The data rate convertor transforms the bursty nature of the raw data into a constant stream. The digital signal has been transmitted using symmetrical differential PSK (SDPSK), also known as 2–4 PSK, where logic values of 1 or 0 are represented by phase changes of plus or minus 90° respectively,

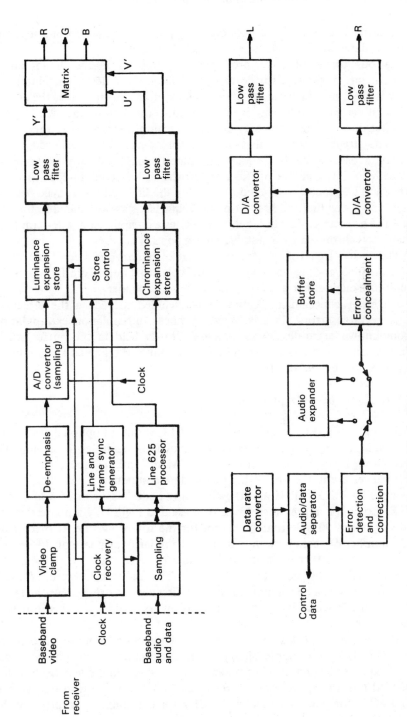

Fig. 7.12 Block diagram of C-MAC (Packet) decoder.

thus using two out of the possible four phases. After reformatting, the control data can be gated from the audio bit stream.

Various forms of error protection are applied to the digital audio component. A single error correction, double error detection, Hamming code is used, plus interleaving of each block of 751 bits to reduce the effects of burst errors. Interleaving is not applied to the data in line 625.

After interleaving, an energy dispersal technique is applied to randomise the data stream and minimise spectrum spreading. This is achieved by the use of a pseudo-random binary sequence (PRBS) generator, running at 20.25 Mb/s. The output of this generator is added, Modulo-2, to the data stream. The generator is gated out during the line and frame sync periods and is so timed that the first bit of the sequence is always added to bit 8 of line 1. The receiver decoder therefore carries a complimentary de-scrambler that is synchronised by a suitable gating control code.

The sound signal may have been near-instantaneously companded, so that a control code is included in the multiplex that automatically routes the audio, through either a linear channel or an expander.

An error concealment technique is included to reduce the effect of uncorrectable errors. This is based on linear interpolation/extrapolation from known accurate values that are already held in the buffer store. Finally, audio outputs that may be stereo, mono or bilingual are produced by D/A conversion and filtering.

D2-MAC Decoder (Digit 2000)

This single chip decoder, shown in Fig. 7.13, forms an optional feature of the Digit 2000 system TV receivers (7), extending facilities to Satellite TV reception. D2-MAC is a sub-set of C-MAC and, as such, the major differences are in the processing of the data components. The data rate is reduced to 10.125 Mb/s and the signalling code is a form of *duo-binary* PSK. Logic 0 is represented by zero phase and logic 1 by a phase shift of plus or minus 90° alternately, except for the following rule. If there is an odd number of zeros betwen two ones, the polarity or phase of the second one is reversed. If there is an even number of zeros, then the phase of the second one is the same as that of the first. The frequency spectrum for this code format has a dc component, but the first zero crossing occurs at 1/2T, instead of 1/T which is common for many binary codes and is therefore half the bandwidth.

The first stage of this chip converts the Gray code from the video codec unit (VCU) into a simple binary code, the luminance component being received as 8 bits and the chrominance as 4 + 4 bits multiplexed. These are clocked into the appropriate stores at 20.25 MHz, to be read out later at 13.5 MHz for luminance and 6.75 MHz for chrominance to produce the correct time expansion.

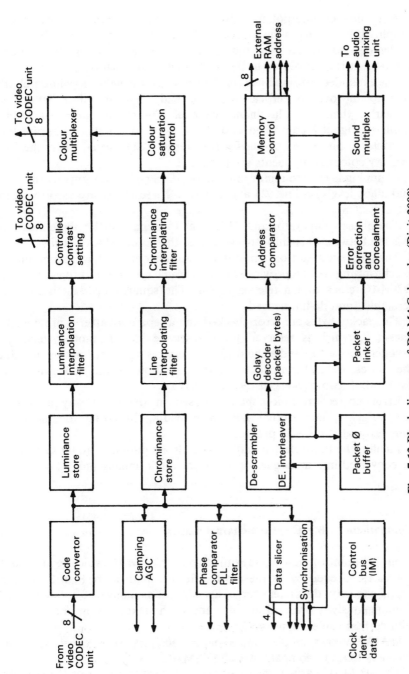

Fig. 7.13 Block diagram of D2-MAC decoder (Digit 2000).

A luminance interpolating digital filter is fabricated on the chip to simplify the filtering in the VCU after D/A conversion. The contrast setting is digitally controlled over the serial control bus (IM or 1^2C). The Y component is then passed to the VCU matrixing circuit.

The expanded chrominance components pass through a line interpolating filter to improve the vertical resolution and, then through a digital filter similar to that in the luminance channel. The colour saturation level is similarly digitally controlled and the 4 + 4 chrominance bits are then multiplexed on to a 4 line bus for transfer to VCU matrix, where the RGB primary colour components are produced.

Clamping is used to remove the energy dispersal waveform and the white and black levels on line 624 are sampled. These values are used to set the AGC system.

The phase comparator and PLL filter form part of the phase lock loop existing between this chip, the VCU and the master clock circuit.

The data slicer section removes any noise from the data stream and the synchronising section provides line and frame sync outputs together with a D2-MAC clock and a teletext signal. The squared up data is then de-scrambled and de-interleaved.

The packet bytes are error-checked and a separate buffer is provided for packet 0, which is reserved for the service identification system. The decoder is able to handle the different services automatically by decoding the address field of the packet header. The 2 bits of the packet linker are used to link two packets of the same service.

Error correction/concealment is used in the same way as for C-MAC/Packet, an external DRAM providing buffering for the sound samples whilst this is in progress.

Finally, the four selected audio channels are output as 16 bit samples, to be converted into analogue format in the audio mixing unit.

Comparison of MAC system characteristics

From the point of view of transmitter power, C-MAC which requires only one carrier frequency, is the most efficient. With more than 3 Mb/s of digital capacity, the system can easily support as many as 8 audio channels simultaneously. However the wide bandwidth makes this unsuitable for use with current cable TV networks.

D-MAC allows simpler and cheaper receivers to be developed, with decoders that are compatible with D2-MAC.

The D2-MAC variant has the same basic characteristics as D-MAC, but with only half the data capacity. However, using current technology, it is the only member of the family, whose signal can easily be recorded on a domestic video recorder.

7.5 HIGH DEFINITION TELEVISION DEVELOPMENTS

It is shown in the literature (8) that the MAC concept has significant scope to produce even higher-definition television. By the use of suitable field stores, the 625 line interlaced format can easily be transformed into 625 lines sequentially scanned. By displaying each frame twice in the normal frame period, the scanning system could be extended to produce a 1250 line 50 frames per second system.

By sacrificing some of the data period in each line and some of the lines in each field blanking period to additional luminance and chrominance information, an Extended MAC (E-MAC) system can be produced. The image extension can be used to provide wide screen displays with an aspect ratio of 16:9.

Using such techniques, it is claimed that the image resolution is limited only by the picture tube itself. Additionally, such a higher-definition TV service can be achieved in an evolutionary way as technology develops.

To provide an HDTV service with a resolution greater than 1000 lines and an aspect ratio of 16:9 requires a base bandwidth in the order of 30 MHz. This automatically precludes the use of such a system as a terrestrial service. The only spectrum available for such a wideband service lies in the Ku and Ka bands of the satellite communications allocations. Furthermore, it has long been recognised that if an HDTV service is to be economically viable, it must be built on the back of an already installed customer/viewer base. This then generates a further dichotomy – whether to introduce the service in a compatible or non-compatible manner. The first viewpoint considers the immediate financial aspects, but in the long run it might be easier and cheaper to forego the advantages and the problems of compatibility that have been so troublesome to the industry in the past. The search for an acceptable single world standard for HDTV has also increased the desire to improve the quality of the current terrestrial systems. These are now well recognised as suffering from unacceptable luminance/chrominance cross-talk and lack of definition.

Super-NTSC

This system, devised by Faroudja Laboratories in the USA, recognises that luminance/chrominance cross-talk arises because of imperfect interleaving of spectral components at the encoder and imperfect separation at the decoder. The source image is produced with a resolution of 1050 lines and progressively scanned at 29.97 Hz frame rate. This is then converted into a sequential (interlaced) scan using a frame store. The luminance component is split into two and the high frequency part accurately comb filtered between 2.3 MHz and 4.2 MHz, the frequencies largely responsible for the cross talk. The chrominance component that covers much the same frequency range is comb filtered in an adaptive manner, the filter coefficient being changed in accordance with the steepness on chrominance transients. Luminance transients that coincide with chrominance ones are also enhanced to improve the horizontal resolution. The basic principles of this concept is shown in Fig. 7.14. A conventional NTSC receiver gains from this pre-processing, but the Super-NTSC receiver also has complimentary comb filtering, plus a frame store which greatly enhances the reproduced images.

PAL Plus

This concept covers a range of possible improvements that vary from transmitting higher frequency luminance information on a second sub-carrier to the applications of adaptive comb filtering much in the manner of the Faroudja system. I-PAL or improved PAL transmits full bandwidth luminance on line n with a low pass filtered (3 MHz) luminance component on line n + 1. QAM chrominance components plus the normal colour burst are included in the multiplex. This system suffers from problems due to phase errors and so the filtering system is modified. This gives rise to the I-PAL-M system.

Q-PAL or Quality PAL, utilises full bandwidth luminance components on all lines. Adaptive three-dimensional filtering is used at both the encoder and decoder to achieve better spectral separation and hence lower cross talk.

HD-MAC

The HD-MAC signal (1250/50/2:1) encoding starts by reducing the number of lines by a factor of 2, to ensure that the signal fits into a standard 6 MHz, 625 line MAC channel and is compatible with a standard MAC receiver. Then by including a digital assistance signal (DATV), that is transmitted in the field blanking interval, the HD-MAC receiver is able to regenerate the missing lines.

In order to ensure that the 25 MHz bandwidth vision signal fits into a standard MAC channel the bandwith has to be compressed by a factor of 4. Fig. 7.15 shows how this is achieved for the luminance signal using ÷ 4 sub-sampling.

The static and moving area digital signals are filtered and then sub-sampled. The two interpolators then reconstruct the signal from the

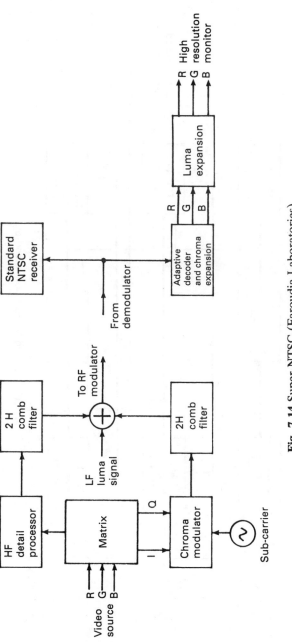

Fig. 7.14 Super-NTSC (Faroudja Laboratories).

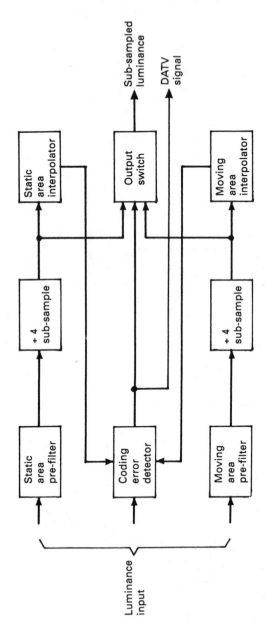

Fig. 7.15 Bandwidth reduction encoding.

reduced sample set. Both versions are then compared with the original input to detect which produces the least error. This is then gated to the output, whilst the minimum error component is used to generate the DATV signal. Thus the encoding is based on an adaptive process that depends upon the motional changes in the image. To take into account motion at different rates, information is derived over 4, 2 or 1 field periods to represent stationary, slow or fast changes respectively. The general principle is shown in Fig. 7.16 where the DATV signal indicates to the receiver not only the degree of motion but also its rate of change.

HDB-MAC

Basically, this wide screen (16:9) system consists of a modified standard B-MAC encoder with extended bandwidth filters (10.5 MHz) to cater for the high definition input signals, together with a pre-processor stage. Using horizontal and vertical filtering, this stage produces spectrum folding at 7 MHz. The encoder then compresses this and adds the other signal components to provide the combined transmission bandwidth of 10.5 MHz.

At the receiver, a standard B-MAC decoder removes the folded component, selects the central 4:3 aspect ratio section and then converts the high-definition signal into a standard NTSC signal.

For high-definition applications, the HDB-MAC receiver decoder processes the same signal. But in this case as a 16:9 aspect ratio, with 525 line sequential scanning at 59.94 Hz field rate using a field store convertor. The system provides vertical and horizontal resolutions of about 900 lines, somewhat below that normally accepted as HDTV.

Hi-Vision or MUSE

This system, known as MUSE (Multiple Sub-NYQUIST Sampling Encoding), is derived from the high-definition TV system developed by the Japan Broadcasting Corporation (NHK) Technical Research Laboratories. It is used by the Sony Corporation, also of Japan, as the basis of the high-definition video system (HDVS) used in video production facilities, producing an image quality that is the equal of 35 mm film but at a very much lower cost. The basic parameters of the system are:

Scanning lines per frame	1125
Active lines per frame	1035
Field frequency	60 Hz
Line frequency	33.75 kHz
Interlace	2 : 1
Aspect ratio	16:9
Video bandwidth	30 MHz

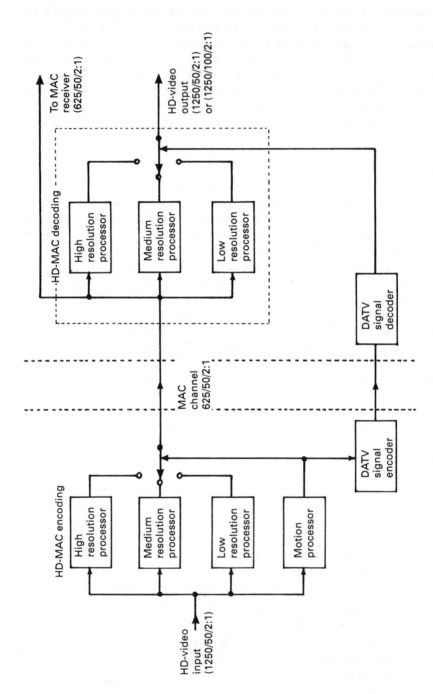

Fig. 7.16 Compatible HD-MAC processing.

As such, this system is not suitable for transmission through a single satellite channel in Ku Band. The bandwidth might be available in Ka Band, but the higher space link noise level demands a higher signal power output at the satellite transponder to maintain an acceptable S/N ratio, thus increasing the demands on the spaceborne HPA and power supply.

By using bandwidth reduction techniques, NHK engineers have succeeded in reducing the baseband requirement to about 8 MHz, for transmission over current Ku Band links (9,10). This has been achieved by using a combination of analogue and digital processing.

The chrominance signal is time-compressed by a factor of 4:1 and the colour difference signals transmitted line sequentially. This information is time-multiplexed along with control and sync information during the horizontal line blanking interval. The luminance component is not compressed and is transmitted in interlaced form. The line multiplex is shown in Fig. 7.17, the three intervals representing 12, 94 and 374 sampling points at the 16.2 MHz rate.

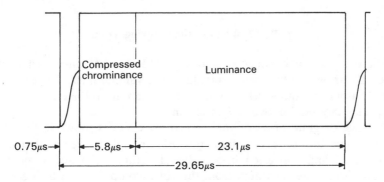

Fig. 7.17 Line multiplex format.

The bandwidth reduction is mainly achieved by the use of spatial and temporal filtering, sampling and sub-sampling. The video signal is sampled at the high rate of 64.8 MHz, which produces 1920 sample points per line. These are stored in a four-field memory and then resampled at the lower rate of 16.2 MHz. This results in the selection of every fourth pixel data. The field sub-sampling pattern and sequence is shown in Fig. 7.18.

At the same time, the image data is compared inter-frame and intra-frame to detect minor movement within a frame. From this, a motion compensation signal is generated. In addition, camera operations such as pan and zoom can create whole image movements which can be compensated for by the application of signals known as *motion vectors*. The motion compensating signals are then multiplexed into the vertical blanking interval for control purposes.

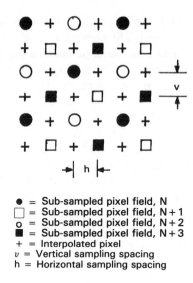

● = Sub-sampled pixel field, N
□ = Sub-sampled pixel field, N + 1
o = Sub-sampled pixel field, N + 2
■ = Sub-sampled pixel field, N + 3
+ = Interpolated pixel
v = Vertical sampling spacing
h = Horizontal sampling spacing

Fig. 7.18 Raster sub-sampling pattern.

The stereo sound channel is sampled at 32 KHz and transmitted as four-phase DPSK within the vertical blanking interval multiplex. The composite signal has a baseband width of about 8 MHz and this is then transmitted by frequency modulation over the space link.

A block diagram of a notional receiver decoder is shown in Fig. 7.19. After conventional processing, the FM signal is demodulated, de-multiplexed and converted into a digital format. The audio, sync and control signals are separated from the video components, which are two-dimensionally filtered and then stored in the four-field memory, with a capacity of about 10 Mbits. The digital data is then clocked out and motionally corrected, the missing pixels being regenerated by intra-frame and inter-frame interpolation.

The colour difference signals are decoded and the Y, R–Y and B–Y components restored to analogue format. A conventional matrix then recreates the primary RGB signals.

This system represents a revolutionary approach to higher definition TV via satellite. Its adoption means a complete re-equipment at the studio, together with new receivers. If MUSE to NTSC, SECAM or PAL convertors are used domestically, the image improvement of MUSE is lost when the converted signal is input to a standard terrestrial system receiver.

Digital HDTV Systems

Throughout the 1980s the American Federal Communications Commission (FCC) insisted that any new HDTV service should be compatible with the

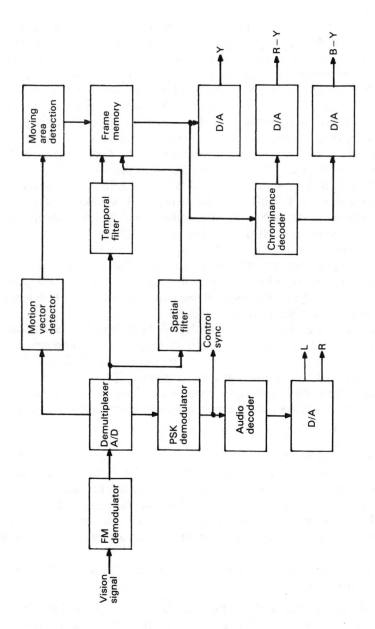

Fig. 7.19 Notional MUSE decoder.

NTSC receivers currently in use. This led to various ingenious proposals, many of which used augmentation channels in which the necessary extra information would be transmitted. Then in early 1990, the FCC decided that this approach was wasteful of frequency spectrum and decreed that a *simultcast* technique would be used. This will allow a station to transmit HDTV in parallel with its current NTSC service using an alternative 6 MHz channel. Compatibility is thus no longer a constraint.

On any broadcast network, transmission frequencies can only be re-used if mutual channel interference is avoided and this feature gives rise to the term '*taboo* channels'. This interference arises primarily due to the power distribution throughout each channel being particularly high in the region of the carrier and sub-carrier frequencies. If a more uniform power distribution can be achieved, then some of the taboo channels become available for the new HDTV service.

This further non-compatible change allows an HDTV service to be lauched without disenfranchising the current NTSC viewer. Over a period of time the current NTSC system will atrophy and viewers will gradually change over to the new system, allowing the present carrier frequencies to be allocated to other services.

To transmit an HDTV signal with a 30 MHz base bandwidth within a 6 MHz channel using analogue methods requires bandwidth compression techniques which to date have not been developed. Therefore the only alternative is to use digital signal processing (DSP).

The bandwidth required to process the luminance component of a 625/25/2:1 image using a 4:3 aspect ratio can be calculated as follows:

$$625 \times 625 \times 4/3 \times 1/2 \times 25 = 6.51 \text{ MHz}.$$

Use of a sampling frequency of 13.5 MHz and a resolution of 8 bits per sample yields a bit rate of 108 Mbit/s. Inclusion of the two chrominance components, each with a resolution of 4 bits per sample, adds a further 108 Mbit/s to give a total bit rate of 216 Mbit/s. At the other end of the scale it can be shown in a similar way that the total bit rate for an HDTV image using the European 1250/25/2:1 format and with a 16:9 aspect ratio yields a total bit rate in excess of 1.1 Gbit/s.

DSP provides the key to significant bit rate reductions which can be carried out in several stages.

(1) By making use of the redundancy that occurs in most images where large areas do not change significantly from frame to frame. This allows a form of differential PCM (DPCM) to be employed.

(2) By manipulating the image pixel data using a two-dimensional (2D) transform technique, followed by scanning the transformed matrix in a manner that produces long runs of near zero values. (See H.261 codecs.)

(3) By using adaptive quantisation of the matrix coefficients to reduce the number of values that need to be transmitted.

(4) By adopting a form of run-length coding where, for example, the values 7,7,7,7 would be coded 7,4 (i.e. 7 four times).

(5) Recently developed modulation schemes that allow each transmitted symbol to represent several binary digits, for example digital 16 quadrature amplitude modulation (16 QAM) based on the PSK system but using 8 vectors spaced by 45°. Since each may occupy one of two levels this provides 16 unique code vectors to generate 4 bits per transmitted symbol. However, relative to bi-phase PSK, reducing the vector spacing from 180° to 45° represents a noise penalty of 6 dB. If technical developments continue along this path, 256 QAM will soon allow each transmitted vector to represent 8 bits per symbol.

Because digital signals are generally more robust than analogue ones under noise conditions, lower transmission power can be used. It is generally agreed that a C/N ratio of 45 dB is necessary to achieve good image quality in an analogue system. With current technology, similar image quality can be achieved with a C/N ratio of around 20 dB, and with a more uniform spectral power distribution.

As an indication of the magnitude of the effort being applied to bit rate reduction, several committees under the auspices of the CCITT (International Telegraphy and Telephony Consultative Committee) involving also the computer industry are very active. These operate under the titles MPEG (Motion Picture Experts Group), JPEG (Joint Photographic Experts Group), and JBIG (Joint Bi-level Image Group (Black and White Imagery).

The following two North American proposals are described because these exemplify the way in which single worldwide digital HDTV standard might develop.

General Instrument Corp, DIGICIPHER System (12)
This all digital system has been designed for transmission within a 6 MHz channel and is therefore suitable for terrestrial as well as satellite distribution. It can operate on relatively low power using small antennas, and a C/N ratio of less than 20 dB will produce error-free reception. In addition the system is taboo channel friendly. The coding system generates a digital data stream multiplex that includes vision, sound and teletext type data. Therefore neither a sound carrier or chrominance sub-carrier is necessary for its transmission. Using digital 16-QAM with 4 carrier phases each with 4 permitted amplitudes produces a peak to average power variation of only 5 dB.

The source image is obtained from R, G, B inputs that are matrixed to produce Y, U and V components with luminance and chrominance bandwidths, or 22 MHz and 5.5 MHz respectively. The image format is based on 1050 lines at a field rate of 59.94 Hz, with 2:1 interlace and 16:9 aspect ratio. The sampling rate is 51.8 MHz with 8 bit resolution for both luma and chroma components. This video bit stream is then compressed to provide a bit rate of 13.83 Mbit/s.

Four audio channels of compact disc quality are provided for by sampling at 44.05 KHz to give a total audio bit rate of 1.76 Mbit/s.

These bit streams are then multiplexed with data and text (126 Kbit/s) and system control data (126 Kbit/s) to give a total effective bit rate of 15.84 Mbit/s. When this is processed through the FEC circuit using a Reed-Solomon (154,130) code the final transmission bit rate rises to 19.43 Mbit/s. Using 16-QAM with its 4 bits per symbol, the final baseband signal becomes 4.86 MHz.

The video signal processing is carried out in five stages:

(1) Chrominance processing
(2) Discrete cosine transform (DCT), (see H.261 codecs)
(3) Adaptive quantisation
(4) Variable length coding (Huffman)
(5) Motion compensation and estimation.

Chrominance information is compressed by the use of decimation filters that averages pixels in groups, four horizontally and two vertically. The luma (Y) signal bypasses this stage before being multiplexed with the processed chroma component one block (8 × 8 pixels) at a time. This multiplex is then passed through the DCT stage with forward (encoder) and reverse (receiver decoder) transformations as follows:

$$F(u,v) = \frac{4C(u)C(v)}{N^2} \sum_{i=0}^{N-1} \sum_{j=0}^{N-1} f(i,j)\, CosACosB$$

$$f(i,j) = \sum_{u=0}^{N-1} \sum_{v=0}^{N-1} C(u)C(v)\, F(u,v)\, CosACosB$$

where $A = \dfrac{(2i + 1)u\pi}{2N}$, $B = \dfrac{(2j + 1)v\pi}{2N}$, i,j & u,v = 0,1, . . . N − 1.

$C(u)$ and $C(v)$ = $1/\sqrt{2}$ for u = 0
and 1 for v = 1,2, . . . N − 1, and
N is the horizontal and vertical block size.

This DCT coder operates with a form of DPCM input. The information for the next frame is predicted and compared with the current image data, and only the differences DCT coded. A good predictor for this is simply the previous frame. The output forms an 8 × 8 matrix of coefficients of which the element 0,0 has a value that is twice the average of the 64 elements. It thus represents the DC energy in the block. The horizontal elements represent increasing horizontal frequencies, and the vertical elements represent increasing vertical frequencies present in the image. The elements along the diagonals thus represent the energy in the diagnonal frequencies. Generally the magnitude of these higher order elements rapidly tend towards very small or zero values, since images only rarely contain significant diagonal information.

The following stage of quantisation or normalisation sets all the small values to zero and truncates certain coefficients in an adaptive manner according to a look-up table stored in ROM.

The matrix of elements is then scanned in a manner that leads to long runs of similar values so that an amplitude/run-length coding can be applied. This takes the form of a modified Huffman code using a two-dimensional code book. If it is detected that a block ends with a long run of zeros, an *end-of-block* code can be appended after the last non-zero value to compress the data stream further.

Motion estimation/compensation is applied to compensate for the errors introduced by temporal image compression. This is achieved by comparing images on a frame by frame basis to detect moving areas and predict how this will appear in the next frame. From this data, motion vectors can be derived for transmission to the decoder. In order to reduce the information needed to describe the motion, estimation is performed on a block matching basis using blocks of 32 × 16 pixels, (a superblock). Since this dimension is compatible with the four-times horizontal and two-times vertical sub-sampling of the chroma signal, a single vector can be used to describe the movement of both components. In this way, the bit rate overhead needed to describe motion is just 9 bits per superblock or about 0.018 bits per pixel.

The variable bit rate output from the Huffman coder needs to be matched to the constant bit rate required for transmission. This is achieved by the use of a buffer memory in both encoder and decoder. This store is capable of holding data for 1 frame ± 1 field variation. The memory level, which is continually monitored, is used to control the adaptive quantisation process to avoid under- and over-flow of the memory.

The 15 KHz baseband audio signal is pre-emphasised, sampled at 44.056 KHz and quantised to 15 bits resolution. This is then instantaneously μ-law companded to 10 bits. The transmission multiplex allows for four such audio channels. Provision is also made to include text type data in 4 × 9600 baud channels.

The control, data, audio, video and sync bit streams are multiplexed in the manner shown in Fig. 7.20. Video lines 2 through to 1050 contain 4 control bits, 4 data bits, 56 audio bits and 440 video bits. For line 1, which

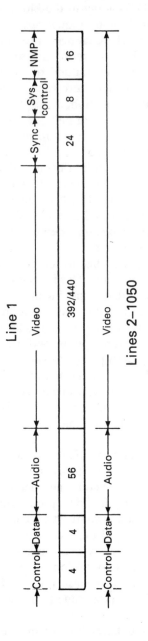

Fig. 7.20 Digital multiplex format.

is chiefly associated with system synchronism, the last 48 bits are allocated to frame sync, system control and next macro-block position (NMP) (macroblock = 256 × 16 pixels). Once the receiver decoder clock has synchronised, the 24 bits in line 1 provides frame sync. The last 16 bits (NMP) are used to support user channel changes and system recovery from error conditions. This is achieved by using this data to signal the number of bits from the end of the NMP field to the beginning of the next macro-block.

The digital multiplex is then FEC coded using a Reed-Solomon (154,130) code before being applied to the QAM transmission modulation system that reduces the 19.42 Mbit/s bit rate to a bandwidth of 4.86 MHz.

Zenith Electronics Corp/American Telephone & Telegraph Corp, (AT&T).
Spectrum Compatible Digital System (13)
Like the Digicipher system with which it has a number of similarities, this all-digital concept also meets the FCC requirements for an HDTV system. As proposed, this system is adaptable for any current means of television signal delivery.

The image input source is gamma corrected R, G, B signals that are matrixed to provide Y, U, and V components. These in turn are bandwidth limited to 34 MHz for luma and 17 MHz for chroma. The image format is based on 787.5 lines per frame and 59.94 frames per second, displayed as 1575 lines progressive or sequentially scanned every 1/29.97 seconds, with an aspect ratio of 16:9. This produces a line scan frequency of about 47.20275 KHz, or exactly three times that of standard NTSC. The image structure provides a high Kell factor that is in the order of 0.9. Conversion of this image format to standard NTSC requires interpolation of only 4:3 horizontally and 3:2 vertically. The digital sampling frequency is 1596 times line frequency or approximately 75.3356 MHz.

Motion compensated transform coding is used to remove the temporal (inter-frame) and spatial (intra-frame) redundancy. Temporal redundancy is removed by estimating the motion of objects from frame to frame using a block matching technique, whilst motion vectors derived from the luminance signal only are used to remove the spatial redundancy.

The principles of encoding are shown in Fig. 7.21. Essentially this predicts how the next frame will appear and, using a decoder similar to that in the receiver, compares this prediction with the actual next frame. Any difference is then applied to the transmitted frame data stream to minimise the error.

The motion estimator compares two successive frames and generates motion vectors for the next frame. These are compressed and stored in the channel buffer for transmission. Each frame is analysed before processing to determine the perceptual importance of each coefficient. Motion vectors and control parameters resulting from forward estimation are then input to

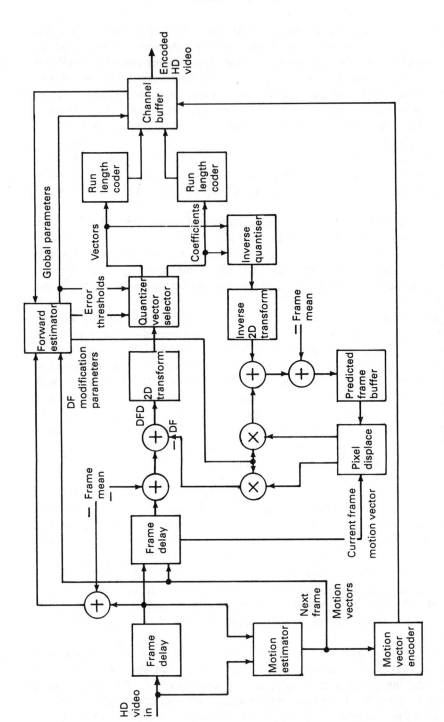

Fig. 7.21 Video encoder.

the encoder loop. This outputs the compressed prediction error to the channel buffer memory, the control parameters being weighted according to the buffer state which is in turn monitored by the forward estimator.

In the predictive loop the data for the new image and the predicted image are coded using an unspecified two-dimensional adaptive transform, the parameters for this coding stage being in part controlled by the forward estimator. Before processing in the prediction loop the input frame has its mean value subtracted in order to produce a *zero-mean* input to maximise the efficiency of the transform stage.

The encoded information that is to be transmitted is also decoded as in the receiver. The pixels of the previously decoded image are displaced (motion compensated) and the result modified by control parameters from the forward estimator. This displaced frame (DF) is subtracted from the input zero-mean frame to produce the displaced frame difference (DFD) and this data stream is then 2D transformed. The coefficient data is quantised and run length coded before being passed to the channel buffer, the luma and chroma difference pixels being coded separately. The total data is then formed into a packet and Reed-Solomon FEC coded (167,147), for transmission. In spite of the variable data rate at its input, the channel buffer output is maintained at about 17 Mbit/s by a control parameter obtained by monitoring the buffer level.

In case of transmission errors or viewer channel changes, the decoder automatically switches to a special mode which allows the image to build up quickly to full quality.

When sound, control and other data is added to the video data stream, the total bit rate rises to 21.5 Mbit/s. This is then used to pulse amplitude modulate the carrier with four discrete levels corresponding to 2 bits/symbol, reducing the symbol rate to about 10.8 Msymbols/s or 5.4 MHz. The transmission uses suppressed carrier, vestigial sideband (VSBSC), with the carrier being positioned half-way down the lower band edge of the 6 MHz channel, 4-VSB being the term being used to describe this form of modulation.

REFERENCES

(1) Carnt, P.S. and Townsend, G.B. (1969) *Colour Television. Vols. 1 and 2.* London: Iliffe Books Ltd.

(2) Hutson, G.H. (1971) *Colour Television Theory.* London: McGraw-Hill.

(3) Weaver, L.E. (1982) *The SECAM Colour Television System.* Tektronix Inc., USA.

(4) Kell, R.D. *et al.* (1934) 'An Experimental Television System'. *Proc. IRE.22*, 1246–1265.

(5) Thorogood, J. ITT Semiconductors Ltd, UK. 'Digit 2000' VLSI Digital TV System. 6250–11–4E. 1985/86. (Intermetall Semiconductors GmbH.)

(6) EBU Technical Document (1984) *Technical Specification for C-MAC/Packet System. SPB 284*. 2nd revised edition. C/D-MAC/Packet Family Specification, Tech. Doc. 3258, 1987.

(7) Thorogood, J. ITT Semiconductors Ltd, UK. DMA 2270. D2-MAC Decoder. 6251–247–5E. 1986. (Intermetall Semiconductors GmbH.)

(8) Annegarn, M.J.J.C. *et al.* (1987) 'HD-MAC: A Step Forward in the Evolution of Television Technology'. *Philips Technical Review* **43**, No. 8.

(9) Takashi, Fujio *et al.* (1985) 'HDTV Transmission Method (MUSE)'. *14th International Television Symposia*, Montreux.

(10) Kimura, E. *et al.* (1984) 'HDTV Broadcast Systems by a Satellite'. *International Broadcasting Convention, IBC '84* (UK).

(11) Slater, J. (1991) *Modern Television Systems – to HDTV and Beyond*. London: Pitman Publishers.

(12) Medress, Dr. M. Video-Cipher Div, General Instruments Corp, California, USA. Private communications to author.

(13) Taylor, J. Zenith Electronics Corp, Illinois, USA. Private communications to author.

Chapter 8

Television Receivers and Distribution Systems

8.1 OUTDOOR ELECTRONICS UNIT

The importance of using low-noise techniques for the first signal processing stages of a system was stressed in chapter 2. Since the signal levels received from a satellite are very small – typically 10 pW or less – the performance of the out-door or head-end unit is particularly critical. This section, the block diagram of which is depicted in Fig. 8.1, is often known as the low noise block-convertor (LNB), because it down-converts blocks of channels to the first IF. The unit consists of several low noise amplifiers (LNA) and a low noise convertor (LNC). Figure 8.1 shows the typical configuration of an LNB for use at 11/12 GHz.

The circuit is typically constructed from thin film micro-strip on a ceramic or similar substrate. The discrete components used are normally of the surface mounting type, for low components loss. The unit is housed in an aluminium casing that is hermetically sealed to provide weather protection, the waveguide input being sealed with a glass or plastic window, the former material being preferred because it is much less prone to embrittlement due to ultraviolet radiation. A waveguide cavity type of bandpass filter may be used to minimise image channel interference and

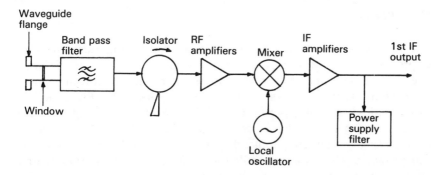

Fig. 8.1 LNB block diagram.

reduce the local oscillator radiation. The latter is further assisted by the use of an isolator, which also helps to improve the impedance matching between the waveguide and the input stage of the LNB. The three or four stages of RF amplification, the mixer and the local oscillator use HEMT GaAs devices, selected for their gain and low noise characteristics. The local oscillator typically employs a dielectric resonator for high-frequency stability. Since the LNB has to drive the main receiver via a significant length of coaxial cable, it is usual to include an IF output circuit that consists of two or three stages of amplification. Power to the LNB is normally provided over the coaxial cable. It is therefore most important that these supplies should be well filtered, to avoid unwanted feedback. By careful selection of the components and by precise tuning of the strip-line circuit elements, it is possible to produce an LNB with a noise factor of 1 dB, equivalent to a noise temperature of 75 K. Fig. 8.2 (courtesy of Mullard Ltd (now Philips Components Ltd)) shows an experimental LNB from around 1985 which clearly shows the waveguide type of input filter. This design was typical of the research being followed during that period to produce low cost units for *direct to home* (DTH) systems. Other techniques tested included fin-line waveguide structures formed by punched sheet metal with the active components mounted directly within the tuned slots. Since that time there have been great developments in GaAs semiconductor technology. Monolithic microwave integrated circuits (MMICs) are now available that considerably reduce the LNB power requirements and heat dissipation, provide a noise factor of less than 1.5 dB and make for a smaller unit. A typical MMIC using HEMT technology consists of RF amplifiers, double balanced mixer, local oscillator less resonator and IF amplifiers. In a typical TVRO application and using a 10 GHz local oscillator frequency, these devices are capable of producing a conversion gain in excess of 60 dB over the band 10.95 GHz to 11.7 GHz. Fig. 8.3 shows a collection of current head-end devices (courtesy of Racal-MESL) together with their important features.

Mixer Stages

The mixer stage is the noisiest in the receiver, and for this reason it should be preceded by some low-noise RF amplification. The noise source is largely due to the non-linearity which creates harmonic distortion, and a conversion loss that represents a noise factor equal to the loss ratio. The oscillator itself can also produce a noise component due to phase jitter. Various mixing circuits have been adopted. These include sub-harmonic mixing using two inverse parallel diodes, with the oscillator operating at a sub-harmonic. This improves oscillator stability, but tends to be noisier and have a greater conversion loss. An alternative method uses two diodes in a balanced mixer configuration. This gives better isolation between the RF stages and the local oscillator.

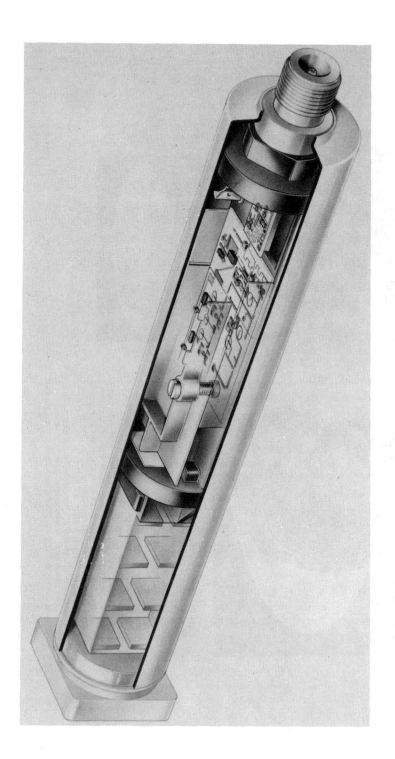

Fig. 8.2 An experimental LNB (courtesy Philips Components Ltd, formerly Mullard Ltd).

1.-Low Noise Polar Converter LNP4505
A fully integrated environmentally sealed unit comprising a state-of-the-art LNB, a low loss ferrite polariser and a high performance broadband feedhorn. V/H polarisation is selected by voltage switching via the RF signal cable.

2.-Feedhorn/Polariser PS45818F
Comprising a broadband feedhorn and polariser the PS45818F is a high quality environmentally sealed unit. By using a metal cover plus 'O' ring seal this innovative design avoids the problems of water ingress and plastic creep.

3.-Broadband Polarisers
Racal-MESL offer a selection of discrete ferrite polarisers, used for linear (V/H) channel selection. Racal's unique technology ensures very high reliability and low insertion loss. A variety of flange fixings are available along with a universal polariser which allows V/H *and* circular depolarisation.

4.-Polariser Inserts "Bobbins" PS4580l
These miniature ferrite polarisers for V/H channel selection are designed for insertion into customer specific feedhorns. For customers with suitable microwave assembly and test capabilities this insert can be procured as a stand alone item.

Fig. 8.3 (Courtesy of Racal–MESL Ltd.)

Voltage Standing Wave Ratio (VSWR) or Return Loss

Impedance mismatching can occur at both ends of the LNB, leading to signal loss and degradation of S/N ratio. It is shown in the literature (1) that this can be quantified by either the VSWR or the return loss.

The reflection coefficient r due to a mismatch is given by:

r = (Power reflected/Power received)$^{0.5}$, or
 = (Voltage reflected/Voltage received) at a load.

Alternatively, in terms of the mismatch impedances:

r = | $(Z_O - Z_L)/(Z_O + Z_L)$| and is complex.
Return Loss = r^2, or in dB,
= $- 10 \log r^2 = - 20 \log r$ dB $\qquad\qquad$ (8.1)
Also VSWR = $(1 + r)/(1 - r)$ $\qquad\qquad\qquad$ (8.2)

G/T Ratio

The equation for the system noise factor that is derived in Appendix 2.1 shows that the antenna with its gain, and the LNB with its noise factor, will have a significant effect on the overall system noise performance. In fact, this is conveniently expressed as a figure-of-merit in the gain/temperature (G/T) ratio. This is perhaps most easily explained by a numerical example.

A 1.2 metre diameter antenna of 65% efficiency has a gain of approximately 41 dBi. It may be pointed towards a signal source with a background noise temperature of 40 K. When used with an LNB that has a noise factor of 1.8 dB, or equivalent noise temperature of 150 K, the total noise temperature is 150 + 40 = 190 K = 22.8 dBK. The system G/T ratio is thus (41 − 22.8) = 18.2 dB/K.

Intermodulation and Non-linear Distortion

Because the LNB is processing a number of carrier frequencies simultaneously, any non-linearity will produce *intermodulation distortion products* – shown in Fig. 8.4 for the two input frequencies f_1 and f_2. Except in the case of very wideband amplifiers, the second order terms $f_2 - f_1$, $2f_1$, $f_2 + f_1$ and $2f_2$ will be well outside of the system pass band. However, the third order terms $2f_1 - f_2$ and $2f_2 - f_1$ can create problems.

Figure 8.5 shows the characteristic for an amplifier with a nominal gain of 10 dB. With increasing input level, the amplifier starts to distort and will eventually saturate. Over the linear part of the characteristic, the gain slope is 1 dB/dB. Assuming that the amplifier is initially saturated due to over driving, the output signal spectrum then contains fundamental plus harmonic components. If the input level is reduced slowly, the harmonic

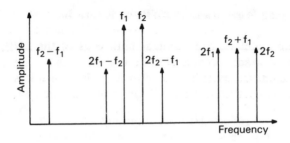

Fig. 8.4 Spectrum of intermodulation products.

components will reduce faster than will the fundamental. In fact, the third order term will reduce with a slope of 3 dB/dB as shown in Fig. 8.5. The slopes of the linear and third order characteristic coincide at a point known as the third order intercept (the theoretical point where the intermodulation product and the signal are of equal amplitude). If this point is known for any given fundamental output power, the level of the third order term can easily be evaluated.

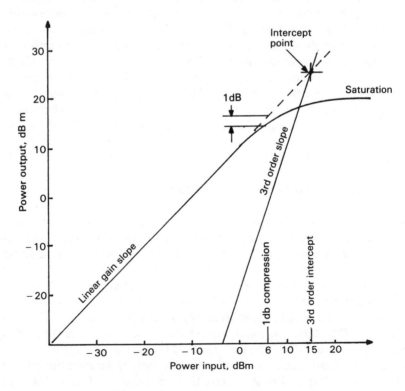

Fig 8.5 Representation of intermodulation products.

Figure 8.5 also shows the 1 dB compression point, a way of representing the degree of non-linear distortion. This point represents the input signal level that produces an output at the fundamental frequency, that is 1 dB less than it would if the amplifier were linear.

For the amplifier characteristic shown in Fig. 8.5, the third order intercept occurs at 15 dBm and the 1 dB compression point at 6 dBm. The higher the value of these two parameters, the lower the degree of non-linearity.

8.2 TV RECEIVERS SAT-TEL MODELS PRK-2 AND PRK-2000

These professional standard receivers (2) are intended for SMATV and CATV applications. Although designed around a common specification, each has features that are specifically related to the special needs of large and small cable systems. Both receivers are future proofed as far as possible against changes of TV signal or encryption formats, by the addition of retrofit plug-in circuits. The block diagram shown as Fig. 8.6 is thus representative of both receivers. Provision is also made for the reception of all FM sound channels using sub-carrier frequencies in the range 5.5 MHz to 8.5 MHz.

Power supply to the LNB is provided from the receiver over the coaxial input cable. The input stage is therefore filtered to prevent the radiation and reception of spurious signals. Both receivers tune over the standard European band of 950 MHz to 1750 MHz for the reception of Ku band signals. The tuned RF amplifier stages are designed to handle input signal levels ranging from −60 dBm to −25 dBm (1 pW to 3 μW) without overloading.

The 2nd IF stages operate at 479.5 MHz and RF tuning and local oscillator frequencies are controlled by frequency synthesis. A remote control tuning facility can be provided via an industry standard RS 232-C interface, with all tuning and control data being stored in a semiconductor memory.

The FM vision signal is processed using a demodulator with a threshold extension to better than 8 dB of C/N ratio. The following video signal de-emphasis is switchable to suit the particular TV signal format being processed.

After filtering and gain control, the video signal is split into two paths. One provides a 5.5 MHz bandwidth output, clamped to remove the energy dispersal component and the other provides a wide band signal (8.5 MHz), unclamped, to feed any decryption decoder that may be needed.

The audio sub-carrier present with the video signal is filtered and buffered before being up-converted to the standard IF of 10.7 MHz, this conversion being controlled via the frequency synthesiser. Demodulation

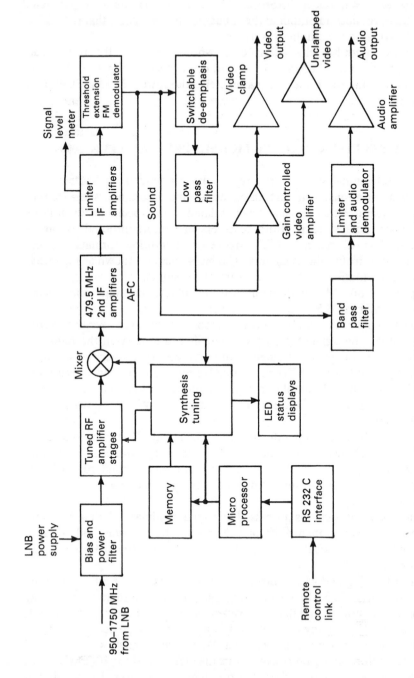

Fig. 8.6 Block diagram of SAT-TEL PRK-2 and PRK-2000 receivers (courtesy of Space Communications (SAT-TEL Ltd)).

then follows standard FM practice but with a range of switchable de-emphasis values (50 μs, 75 μs, J17).

8.3 FREQUENCY SYNTHESIS TUNING

A simplified block diagram of the SP5000 frequency synthesis IC (3) is shown in Fig. 8.7. A programmable counter and prescaler are additions to the basic PLL. These are used to extend the range of operation. The basic reference frequency is derived from the 4 MHz crystal oscillator after division by 1024 in two stages. The receiver local oscillator is thus locked to multiples of 3.90625 KHz.

When a channel number is selected, the control logic converts this into a frequency and the phase comparator causes the local oscillator to be adjusted until the output from the variable divider is equal in phase and frequency to 3.90625 KHz. As the prescaler has divided the local oscillator frequency by 16 to achieve this, the actual tuning steps will be $3.90625 \times 16 = 62.5$ KHz. This frequency is small enough to allow the receiver AFC system to provide final tuning accuracy.

Later ICs in the SP series are equipped for microprocessor control over the standard inter-integrated circuit (I^2C) bus. This consists of two lines forming a ring network to which all the controlled and controlling chips are connected in a master/slave fashion. One line provides for serial data up to 100 kbit/s and the other carries the clock signal.

Direct digital synthesis (DDS) is a technique that is rapidly gaining in popularity. This concept can provide a system with lower phase noise than direct analogue synthesis (DAS) described above and in addition is more compatible with microprocessor control. Essentially this system consists of a register/accumulator type memory, an address bus, a ROM look-up table and a digital to analogue convertor (DAC), all controlled by an accurate clock circuit.

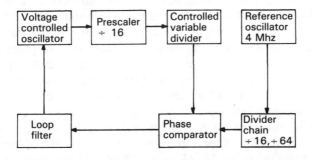

Fig. 8.7 Simplified block diagram of frequency synthesisers (courtesy of Plessey Co. Ltd).

The required tuning frequency is entered into the register and this binary number is used to address the ROM whose contents represent the binary code associated with this frequency. When this output is converted into an analogue form it represents the tuning frequency. As an example, the output frequency is given by:

$$f_{out} = \frac{f_{clock}}{2^n} \times \text{Input data value}$$

where n is the number of address bits.

If the system uses a 16 bit address sytem and a clock frequency of 327.68 MHz, then the system will tune in steps of

$$\frac{327.68 \times 10^6}{2^{16}} = 5 \times 10^3 \text{ Hz}$$

up to the clock frequency.

Higher frequencies can be generated by mixing the output step values in a frequency convertor stage.

8.4 TV RECEIVER DRAKE SERIES II MODELS (1024, 1224, 1424)

This series of North American receivers (4) is designed for both C and Ku band operation to provide *direct-to-home* (DTH) television services. Although the receivers described here and shown in Fig. 8.8 are for the NTSC television format, versions are available for European PAL and SECAM. All receivers are microprocessor controlled for tuning and antenna positioning via alternative forms of remote control, a conventional infra-red system or a UHF transmitter designed to give a much greater range of control. In addition, on-screen graphics are provided to aid the setting up and tuning of the receivers. The 1st IF tuning range extends from 950 MHz to 1450 MHz and the tuned RF amplifier section is capable of handling input signal levels ranging from −60 dBm to −25 dBm without overload. Because the LNB is powered via the coaxial cable feed from the receiver, the input stage is fully filtered to prevent spurious interference.

Tuning for the RF and local oscillator stages is microprocessor controlled, with user tuning data being held in a semiconductor memory. The microprocessor also provides outputs to drive the front panel status lamps and together with a character generator provides the on-screen graphic signals that are gated through the video amplifier stage.

The 2nd IF amplifier stage, which operates at 140 MHz, uses surface acoustic wave (SAW) filters to provide much of the selectivity for switchable bandwidths of 18 MHz or 27 MHz. These stages which use conventional gain control (which also provides the drive for the signal level

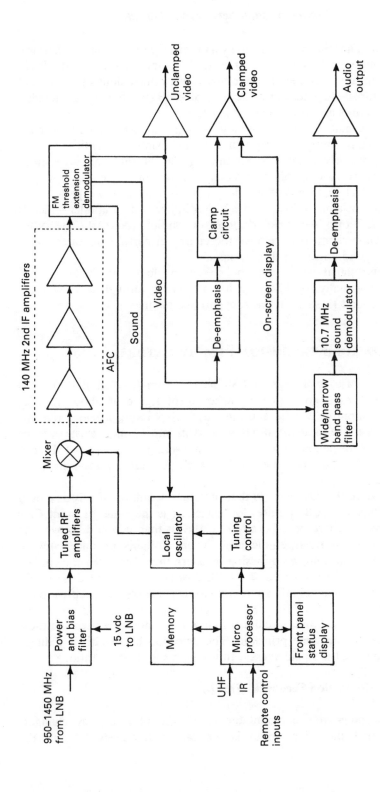

Fig. 8.8 Block diagram of Drake Series II Receivers (1024, 1224 and 1424) (courtesy of R. L. Drake Co).

indicator) and limiting are followed by a unique FM demodulator. This is a discrete component balanced circuit using two pairs of hot carrier diodes in charge pump configuration. This is effectively a voltage doubler circuit and achieves a threshold extension to better than 7 dB of C/N ratio. Both video and AFC circuits are driven from this stage.

Provision is made for the addition of a switchable terrestrial interference (TI) filter as a retrofit plug-in to reduce the pass bandwidth of the IF stage. This is particularly helpful in minimising the TI that can arise from terrestrial microwave telephony circuits operating in C band.

Following demodulation, the video signal is split into two paths. One provides an unclamped drive to the VideoCipher II descrambler which is virtually the North American *de facto* encryption standard. The other drive, which is clamped to remove the energy dispersal component, also carries the gated on-screen display signal.

The audio signal which may be stereo, analogue or digital, is up-converted nominally to 10.7 MHz for processing through a SAW filter controlled sound IF stage in the conventional manner.

8.5   THRESHOLD EXTENSION DEMODULATORS

Part of the noise advantage of FM derives from the fact that the FM demodulator has a noise power improvement figure of about 15 dBs, between the input C/N ratio and the output S/N ratio, provided that the input level is above some *threshold* value. This improvement exists over the linear part of the demodulator characteristic that has a slope of 1 dB/dB. When the C/N ratio falls below this level, FM rapidly loses its noise advantage.

It is shown in the literature (5) that the usual definition of this threshold level is a kind of 1 dB compression effect. It is the value of input C/N ratio at which the output S/N ratio is 1 dB less than it would be over the linear region. Below the threshold level, the onset of noise produces *clicks* in audio systems and *sparklies* on vision. Modern threshold extension circuits have reduced this level from about 14 dBs down to about 5 to 6 dBs of C/N ratio, thus providing a significant margin to relax the system specification in other areas.

Basically, there are three circuit configurations that can be used and these all function on the same principle. The detection bandwidth is reduced without reducing the signal energy, thus enhancing the S/N ratio.

Frequency Modulation Feedback (FMFB)

Figure 8.9 shows this demodulator as a type of frequency changer. Assuming that the mixer input is the carrier frequency f_c, plus the

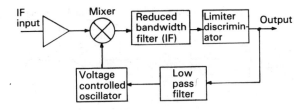

Fig. 8.9 Frequency modulated feedback demodulator.

instantaneous deviation f_d, $(f_c + f_d)$, the output from the descriminator is a voltage v, proportional to f_d. When this is filtered and applied to the VCO with a nominal frequency of f_o, it will produce a change of frequency, say kf_d. The second mixer input is thus $f_o + kf_d$. The difference signal at the mixer output is therefore $f_c + f_d - f_o - kf_d$. Now $f_c - f_o$ is a new intermediate frequency f_i, so that the mixer output becomes $f_i + f_d - kf_d = f_i + f_d(1 - k)$. Since k must be less than unity, the deviation has been reduced to allow the use of a narrow bandwidth filter, thus reducing the detection bandwidth and hence the noise power.

Dynamic Tracking Filter Demodulator

The circuit configuration shown in Fig. 8.10 is similar in form to the PLL. However, the *tracking filter* is an active device with a narrow bandwidth and with a centre frequency that is made to track the relatively slowly deviating FM carrier under the feedback control. Since the detection noise bandwidth is just the tracking filter bandwidth, the S/N ratio at the output is improved by about 6 dBs.

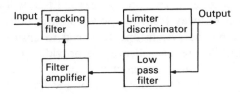

Fig. 8.10 Dynamic tracking filter demodulator.

Phase Lock Loop Demodulator

Figure 8.11 shows how the basic PLL is modified by the addition of a secondary feedback loop filter. Due to the feedback, the VCO is made to follow the deviations of the input FM carrier. Its driving voltage is thus dependent upon the deviation, and so represents the modulation

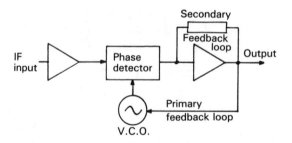

Fig. 8.11 Phase lock loop demodulator.

component. The circuit therefore functions as an FM demodulator. The addition of a narrow bandwidth secondary feedback loop in the form of an active filter reduces the detection bandwidth to improve the S/N ratio.

8.6 TELEVISION DISTRIBUTION TECHNIQUES

Apart from the single user television receive only (TVRO) application, there are two other important distribution concepts: the Satellite Master Antenna TV systems (SMATV), serving a single site such as a hotel or block of flats, and the full community antenna TV system (CATV) or cable system. The latter is already well documented (6) and will only be considered here in a limited way.

SMATV System

Figure 8.12 shows one half of a typical SMATV system, designed for operation with the Fixed Satellite Services (FSS). A line amplifier stage is needed because the antenna site, and hence the LNB, will most probably be more than 50 metres away from the power splitter unit. Due to the extra loading imposed by the system and the need to ensure a high grade of service, the SMATV antenna will need to provide an extra 3 to 4 dBs of gain. For example, if an antenna of 0.8 to 1.2 metres diameter would provide an adequate TVRO service, then for a SMATV system at the same site an antenna of 1.8 to 2.2 metres diameter should be installed. After signal splitting, using an active device, each output can be fed to separate receivers, each tuned for the vision and sound parameters of that transponder channel. Where a channel is carrying encrypted signals, the wideband unclamped video output will be used to drive the necessary decoders. The baseband signals are then remodulated on to separate UHF carriers and combined for distribution to standard UHF TV receivers.

The two LNBs for each of the X and Y linear polarisations can be fed from the same antenna using an ortho-mode transducer. The complete system will be twice as extensive as that depicted in Fig. 8.12.

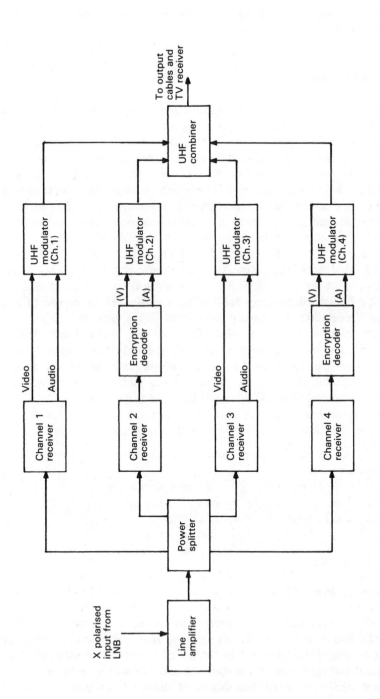

Fig. 8.12 Satellite master antenna TV system.

For the DBS service, the LNB feeds will need to accommodate both left and right-hand circular polarisations.

CATV Considerations

Metallic coaxial cables or optical glass fibres may be used for the new distribution systems, the final choice being a compromise based on the cost effectiveness of the system as a whole. Coaxial cables are leaky to RF, radiating and receiving unwanted signal energy. Although the cable bandwidth may be as high as 500 MHz, certain carrier frequencies have to be avoided (the taboo channels). These are related to the local off-air service and include the actual local carrier frequencies and their images, plus the local oscillator frequencies of receivers tuned to such channels. This produces such a restriction that several parallel cables have to be used to provide spectrum for the required number of channels.

Coaxial cable attenuation increases with frequency, and at 400 MHz may vary between about 3 dB/100 metres for a trunk cable and 15 dB/100 metres for a subscribers drop cable. These values entail the use of line amplifiers (repeaters) about every 750 metres of the network.

The attenuation of optical glass fibre is very much lower, but is dependent upon the fibre technology and the light wavelength in use. Typical losses are less than 2 to 4 dB/km, but may be as low as 0.2 dB/km for special fibres. Signal bandwidths as high as 5 GHz have been reported for operation in the infra-red range of 1.3 to 1.6 μm.

The constructional features and properties of optical fibres is well documented (7), but some of the important advantages are as follows:

(1) Low cost, abundant supply raw material (silica, sand) with falling production costs.
(2) Very much wider bandwidth and lower attenuation.
(3) Glass, being an electrical insulator, does not generate and is not influenced by electromagnetic interference.
(4) Very many fewer repeaters needed in the system.
(5) There is no standing wave/impedance mismatch problem associated with fibre jointing.

Fibre Jointing Methods

There are two basic methods of forming joints between sections of optical fibres. The first is simple. It involves accurate abutment of the two ends and then the sealing of the joint with an ultra-violet light-curing adhesive. Such joints may produce losses of less than 0.5 dB. The major problem of joint losses arise from the small eccentricity of the core and cladding glasses.

Joints made by the alignment of the outer glass diameter may thus be suspect.

An alternative, but more effective, method is provided by the *fusion splicer* (8). Here light is injected into one of the two fibres and extracted from the other. The two ends are then manipulated under microprocessor control, to maximise the light transfer. At this point, an arc is struck between a pair of electrodes that causes the glass to fuse. Joints made in this manner have a loss in the order of 0.05 dB. Figure 8.13 (9) shows this jointing technique very clearly.

The ideal cable system is therefore likely to be a hybrid combination of optical glass fibre for the main trunks, with coaxial cable being used for the subscribers feed.

Network Configuration

The traditional cable systems of the past have been based on the *tree and branch* configuration as depicted in Fig. 8.14(a), where it is difficult to provide for subscriber-to-subscriber interactivity. This concept is therefore giving way to the *switched star* network as indicated in Fig. 8.14(b). By including suitable switches at each system node, complete flexibility is provided for interactive services.

Each switching unit in the system is based on a matrix formed by the incoming lines and the outgoing subscribers feeds. This leads to two possible switching methods: either by switching at baseband, which requires demodulation at the switch and final distribution at baseband, or by remodulation on to a new carrier. The preferred method uses a frequency translator type of switch. The incoming VHF distribution signal is frequency-changed to a UHF carrier for direct feed to a standard TV receiver. Matrix switches are still required at each cross point, but these can be fabricated using GaAs semiconductor technology, in the manner shown by Fig. 8.15. By providing simultaneous negative and positive voltages at V_1 and V_2, a fast switching action is generated, with a loss of less than 0.7 dB in the ON state and an isolation greater than 30 dB when OFF (at 500 MHz).

Frequency Agile Switching

The technique shown in Fig. 8.16 represents one way in which a frequency agile switching system can be implemented (10). The channel switch for TV is based on a matrix formed by the six trunk cables and the three feeds to each subscriber, the cross-point switching and channel selection being performed in two stages, which involves a two-digit program selection. These digits define the wanted channel and cable respectively.

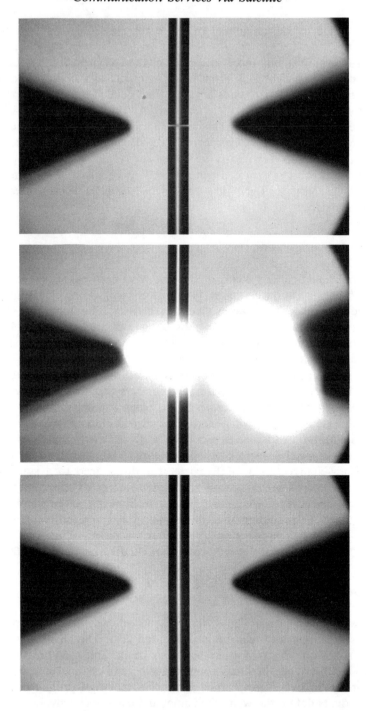

Fig. 8.13 The fusion splicer process (courtesy of BICC Research and Engineering Ltd).

(a)

(b)

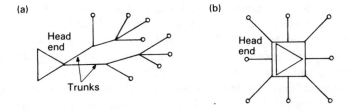

Fig. 8.14 Networks: **(a)** tree and branch; **(b)** star.

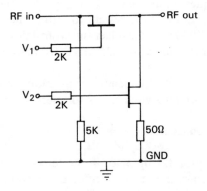

Fig. 8.15 GaAs semiconductor RF switch (courtesy of Plessey Co. Ltd).

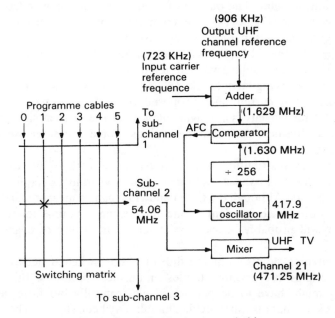

Fig. 8.16 Channel selection switching.

In the example represented in Fig. 8.16, UHF channel 21, with a vision carrier at 471.25 MHz, has been selected. This signal has been delivered to the switch at 54.06 MHz. Within the switch unit, two groups of highly accurate reference frequencies are generated from transmitted pilot signals. These are five square wave switching signals in the range 343 to 723 KHz and three sine waves in the range 906 KHz to 2.38 MHz. The two groups are used as references for the input and output frequencies respectively.

In the example shown, the references 723 KHz and 906 KHz are selected and added to produce one input to the AFC comparator at 1.629 MHz, the local oscillator frequency, divided by 256, providing the second input. The frequency error causes the AFC system to retune the varicap diode tuned local osillator to 417.9 MHz. The mixer thus translates the input VHF signal to the desired channel value of 471.25 MHz, for delivery to the subscribers receiver.

Now that the correct frequency translation has been set up, a semiconductor cross-point switch closes to select the wanted cable.

A broadband cable concept

Due to the deregulation taking place in the telecommunications industry, cable TV operators are now able to offer the customer/subscriber many other services. Bandwidth to provide full duplex interactive telephone type services and television is readily available using optical fibre as the transmission medium. The outline of such a computer controlled system (10) is shown in Fig. 8.17. The computer not only manages system housekeeping, conditional access and customer billing, but also provides an interface to the PTT (British Telecom and Mercury in the UK) for connection to the public switched telephone network (PSTN) or the integrated services digital network (ISDN). The network would be capable of supporting all current and proposed future TV formats, including HDTV. A data capacity is also available for such services as slow scan TV surveillance, business data communications, home computer users, tele-conferencing, etc.

The primary distribution is over a trunk network of optical fibres. Because of the low loss, distribution amplifiers will only be required at the switching points. Although the fibre bandwidth can be measured in GHz, it is inversely distance dependent. In the initial stages, the final connection to the home will probably be via a relatively short length of coaxial drop cable.

The switching points form the hub of a star switched network, each serving perhaps 100 or more homes or business premises. Initially the switching might have to be performed electronically but fibre developments are such that optical switching has already been shown to be a viable proposition. The trunks would probably consists of up to eight fibres, each

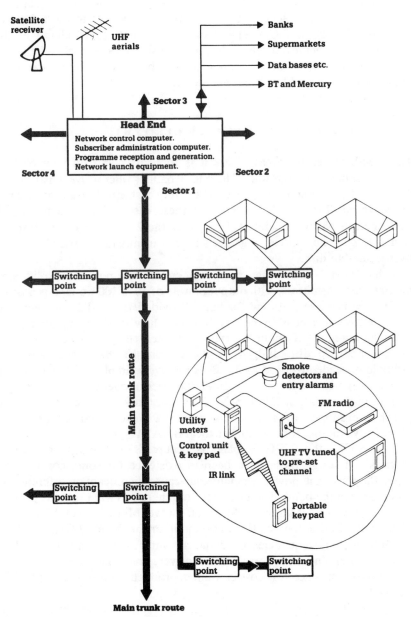

Fig. 8.17 A broadband cable system (courtesy of Rediffusion Engineering British Cable Services Ltd).

carrying up to three TV or 20 radio channels, with provision for reverse communications. The signals to be transmitted would modulate laser light sources and use wavelength division multiplex (WDM) in the ranges of 1300 nm or 1550 nm. Conversion of these signals to baseband at the receiver can be achieved relatively cheaply using PIN or avalanche photo diodes (APD). The latter would generally be preferred because of its higher sensitivity and low noise characteristic.

8.7   AN ALTERNATIVE TO CABLE DELIVERY

There will be certain areas where it is too costly and difficult to bury network cables underground. The Multichannel Microwave Distribution Service (MMDS) might well prove to be a very cost-effective alternative way of delivering TV and data signals in these cases. This system, already developed in North America (11,12), uses the 2.5 GHz band to transmit signals from local sources, together with retransmission of signals received over a satellite link.

It has been shown that by using small, low-powered, solid state microwave transmitters, with output powers of 10 to 100 watts, it is possible to provide coverage beyond 30 kms. Provided that a line-of-sight path exists, adequate reception quality can be achieved with simple parabolic receiving antennas of about 0.5 metre diameter, the gain typically being in the order of 20 dBi. Judged by the current Ku Band standards, the technology needed to produce LNBs is relatively simple and low cost. The LNB acts as a convertor to UHF or VHF carrier frequencies for direct input to standard receivers. Thus MMDS is not only capable of providing distribution support for a difficult section of a conventional network, but is also highly suitable for the smaller local SMATV systems.

In the UK, the 2.5 GHz band is occupied by tropospheric scatter services. Although these have been in existence for some considerable time, they are not power efficient and cause significant interference to other users in the band. The UK operators therefore will probably provide an equivalent service known as Millimetre wave, Multichannel, Multipoint, Video Distribution Service (M^3VDS) in either the 29 or 39 GHz bands.

These higher frequencies have the advantages that antennas will be much smaller and less obtrusive. However, the high level of R & D necessary to produce low-cost consumer products will delay its introduction.

REFERENCES

(1) Hewlett Packard Inc. (1964) *Applications Note No. 16: Waves on Transmission Lines.*

(2) Knowles, J. Space Communications (Sat-Tel) Ltd, Northampton, UK. Private communication to author.

(3) Salter, J. (1983) *"SP5000" Single Chip Frequency Synthesis*. Plessey Semiconductors Ltd (UK).

(4) Jackson, B., R.L. Drake Company, Ohio, USA. Private communication to author.

(5) Beech B. and Moor, S. (1985) *Threshold Extension Techniques*. IBA Report 130/84. Independent Broadcasting Authority UK.

(6) Maynard, J. (1985) *Cable Television*. London: Collins Professional and Technical Books Ltd.

(7) Senior, J. (1985) *Optical Fiber Communications: Principles and Practice* London: Prentice-Hall International Inc.

(8) Andrews, P.V., Grigsby, R. *et al.* (1984) A Portable Self-aligning Fusion Splicer for Single-mode Fibres'. *International Wire and Cable Symposium*. Nevada, USA. (BICC Research and Engineering Ltd, UK.)

(9) Grigsby, R., BICC Research and Engineering Ltd, UK. Private communication to author.

(10) Quinton, K.C., Rediffusion Engineering Ltd (British Cable Services Ltd, UK.) Private communication to author.

(11) Henry, D.G. (1986) DGH Communications Systems Ltd, Ontario, Canada. Private communication to author.

(12) Evans, W.E. (1986) 'MMDS Technology – An International Opportunity'. *IBC'86 Conference Proceedings*. Brighton, UK.

Chapter 9
Television and Radio Audio Channels

9.1 TOWARDS STEREO TV

Throughout the development period of television engineering the sound channel has tended to be neglected, at least as far as receiver design and construction is concerned. Although broadcasters have invariably tried to provide a high-quality signal, set manufacturers have failed to find a significant marketplace for television with hi-fi sound. The cabinet of the typical vision receiver can in no way match the acoustic properties of even a simple design of loudspeaker enclosure.

The big improvements in audio quality provided by magnetic tape systems and the introduction of compact disc players appear to have created a new enthusiasm for better audio quality, plus stereo, to enhance the total viewing experience. This can only be further encouraged by satellite television, where the overspill of national boundaries and the increasing number of polyglot communities produce a demand for parallel bilingual sound channels. Although the Zenith GE (pilot tone) system, as used for terrestrial VHF/FM radio, provides a very satisfactory service, no such similar standard has evolved for use with satellite television.

9.2 THE LEAMING SYSTEM

In this variant of the Zenith GE system, the two audio stereo channels (L and R) are first added and subtracted. The difference signal (L−R) is used to amplitude-modulate a 38 KHz sub-carrier using DBSSC. In order to regenerate an accurate sub-carrier at the receiver, a 19 KHz pilot tone is added, together with the sum signal (L+R), to form the baseband frequency multiplex. This may be companded to improve S/N ratio, before being used to frequency-modulate the final RF carrier. The receiver decoder has to contain a *whistle filter* to remove the beat note between the pilot tone and the line timebase frequency.

9.3 BROADCAST TELEVISION SYSTEM COMMITTEE/MULTICHANNEL TELEVISION SOUND SYSTEM (BTSC/MTS)

This variant of the Zenith GE system uses a pilot tone and sub-carrier locked to the horizontal timebase frequency to avoid the beat note problem. The baseband frequency multiplex is shown in Fig. 9.1. In addition to the sum and difference signal components, a separate audio programme (SAP) channel is provided. This represents a 12 KHz wide audio band, which is frequency modulated on to a sub-carrier at five times the line frequency, both the L–R and SAP channel being companded. The SAP channel may provide an alternative language version of the main programme, or may be completely unrelated to it. In certain cases, a further *professional channel*, may be located at 6.5 times line frequency. This narrow-band (3.4 KHz) channel is provided for *talkback* during outside broadcasts. The decoder phase adjustment can be critical and such errors can impair the channel separation.

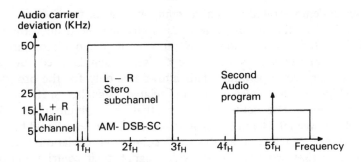

Fig. 9.1 BTSC multi-channel stereo sound spectrum.

9.4 JAPANESE FM-FM SYSTEM

The sub-carrier of this system is locked to the second harmonic of the horizontal timebase. The sub-carrier channel may carry either the (L–R) stereo component or a mono second language signal as frequency modulation. An amplitude-modulated 55 KHz sub-carrier is used to convey to the receiver decoder the necessary control signals for automatic switching between bilingual or stereo operation. The composite baseband multiplex, which is shown in Fig. 9.2, is then used to frequency-modulate the final RF carrier. The system offers good mono/stereo/bilingual compatibility, is

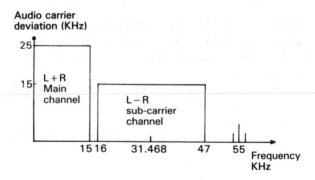

Fig. 9.2 Japanese FM-FM system spectrum.

easy to implement and the decoders are relatively simple. However, the rather wide bandwidth can give rise to adjacent channel interference problems.

9.5 MULTI-CARRIER CHANNEL OPERATION

It has been demonstrated (1) that as many as eight or ten FM sub-carriers can be added to the normal video/sound channel spectrum without any significant effects on the video S/N ratio or the demodulator threshold. Wegener Communications Inc. have devised a standardised sub-carrier band plan for such operation. This allows not only for the programme audio channel, but also for non-related signals, such as data communications. The main criteria are that the FM modulation index should be between 0.14 and 0.18 and that the sub-carriers should be spaced by 180 KHz. Provided that the total sub-carrier deviations are small compared to that produced by the video signal, their contribution to the total bandwidth is insignificant.

Each 180 KHz slot can be allocated to 15 KHz of audio, or further sub-divided for either 7.5 or 3.5 KHz of audio, or even for data, using FSK or QPSK, etc. One Dolby ADM channel can be accommodated in two adjacent 180 KHz slots. For NTSC applications, the sub-carriers are typically disposed between 5.2 and 8.5 MHz. The equivalent distribution on PAL systems lies between 6.3 and 7.94 MHz.

9.6 WEGENER 1600 STEREO SYSTEM

This stereo system is a sub-set of the Wegener band plan. The left (L) and right (R) audio channels each frequency modulate (FM) separate sub-carriers spaced by 180 KHz. These FM sub-carriers are then used to frequency modulate the final RF carrier, a technique often referred to as

FM[2], two commonly used sub-carriers being 7.02 and 7.20 MHz. Companding of both audio channels is employed to improve signal to noise ratio and a sub-carrier deviation of $\pm$ 50 KHz is allowed.

Two different companding standards may be used. PANDA I is a broadband, 2:1 linear system which under certain conditions suffers from system noise. It is therefore only satisfactory for high-level audio signals. PANDA II which is now more common resolves the problem by splitting each audio channel into two at 2.1 KHz and then applying companding with a ratio of 3:1 to each segment.

9.7 WARNER AMEX SYSTEM

This is a dual sub-carrier system, in which sum $(L + R)$ and difference $(L - R)$ audio signals are generated and used to frequency modulate separate sub-carriers. This multiplex is then in turn used to frequency-modulate the main sound carrier.

9.8 WEST GERMAN DUAL CHANNEL STEREO SYSTEM

Unlike the Zenith GE variants that transmit the sum $(L + R)$ and difference $(L - R)$ components within a baseband multiplex, this system transmits the sum signal on the main carrier and additional information on a sub-carrier. This is done because the conventional decoding matrix that is used for sum and difference components, shown in Fig. 9.3(a), typically only achieves about 35 dB of channel separation. This is insufficient for dual language operation. In addition, the noise due to an interfering signal (N), which will affect each channel equally, tends to concentrate in one channel output. This can be explained as follows, assuming that the two inputs are $0.5(R + L) + N$ and $0.5(R - L) + N$ respectively:

For the sum channel:
$$(0.5(R + L) + N) + (0.5(R - L) + N)$$
$$= R/2 + L/2 + N + R/2 - L/2 + N$$
$$= R + 2N \tag{9.1}$$

For the difference channel:
$$(0.5(R + L) + N) - (0.5(R - L) + N)$$
$$= R/2 + L/2 + N - R/2 + L/2 - N$$
$$= L \tag{9.2}$$

The noise is thus shown to have concentrated in the right channel.

By transmitting the stereo signals as $R + L$ and $2R$, and by using the decoding matrix shown in Fig. 9.3(b), the new conditions are as follows:

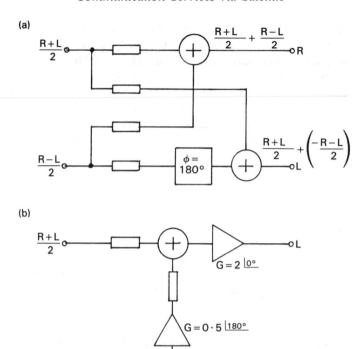

Fig. 9.3 Stereo decoder matrices: **(a)** Zenith system; **(b)** West German system.

For the left channel:
$$2((0.5(R+L)+N) + (-0.5(R+N)))$$
$$= R + L + 2N - R - N$$
$$= L + N \tag{9.3}$$

The right channel is simply:
$$= R + N \tag{9.4}$$

Thus the noise is now equally distributed across the two channels.

This state is achieved by the use of two carriers: one placed at 5.5 MHz as normal, and the second at 5.742 MHz above the vision carrier (Fig. 9.4). This latter may carry a second language. In order to minimise any cross-talk, its level is maintained 7 dB below the level of the main sound carrier. Compatibility with mono receivers is assured because the main carrier conveys the sum (L + R) signal. There are thus three modes of transmission: mono, stereo and bilingual. Mode control is automatically affected by additional signalling on the second carrier. This is frequency-modulated with a deviation of ±2.5 KHz, with an identification signal consisting of a 54.6875 KHz pilot sub-carrier, which is unmodulated for mono transmissions and amplitude-modulated to 50% by tones of 117.5 Hz and 274.1 Hz for stereo and dual language sound respectively. Both the sound

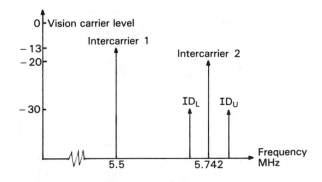

Fig. 9.4 Dual channel spectrum for audio signals.

carriers and the control signals are multiples or sub-multiples of the line scan frequency.

The basic operation of this decoder is explained with the aid of Fig. 9.5. The modulation recovered from the two sound carriers is dematrixed as necessary and then de-emphasised. An ID decoder identifies the particular mode of transmission and generates logic-switching signals. These allow the audio switch unit to provide the appropriate outputs. As most receivers designed for this system have headphone listening facilities, the switch unit provides dual outputs.

The system is relatively economical to implement and has the necessary mono/stereo/bilingual flexibility. Under conditions of adjacent channel interference, however, a buzz on sound and or a patterning on vision can be produced.

9.9   GORIZONT SOUND CHANNELS

This USSR C Band system transmits SECAM TV with the main sound channel frequency-modulated and located 7 MHz above the vision carrier. It also carries a radio programme on a sub-carrier of 7.5 MHz. The TV sound channel, which has a base band extending to 10 KHz, is non-linearly companded. The low-level audio components thus produce almost the same deviation as do the high-level ones. An 11 KHz pilot tone is added to convey the control information to the receiver expander, the pilot tone being amplitude-modulated at syllabic rate. Figure 9.6 shows how the received audio component is processed. The audio and pilot tone are separated using filters; the pilot tone is demodulated and its varying dc voltage used to control the gain of a voltage-controlled amplifier, thus restoring the original dynamic range. An 11 KHz notch filter is needed to prevent unwanted whistles due to the pilot tone. The separation filters are rather critical. Any high-frequency audio component reaching the pilot tone detector will

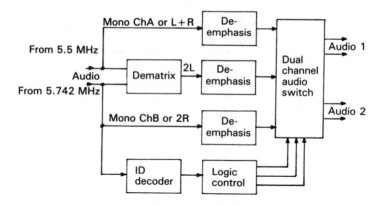

Fig. 9.5 Block diagram of West German stereo system decoder.

generate a false control, allowing accentuation of sibilants ('essing'). The time constant circuit is also fairly critical. If this is too long, the expansion will not be sufficient; if too short, the audio level will fluctuate.

9.10 DOLBY ADAPTIVE DELTA MODULATION DIGITAL AUDIO SYSTEM

Delta modulation (DM) uses only 1 bit per sample to indicate whether the analogue signal is increasing or decreasing in amplitude. This effective bit rate reduction technique allows the use of a higher sampling frequency, which in turn leads to a simpler filter arrangement in the decoder, without the risk of aliasing. Unlike PCM, a single bit in error produces the same signal effect wherever it occurs. When a bit error is detected in a DM system, the introduction of an opposite polarity bit will reduce the audible effect to practically zero. The only major disadvantage is that an overload can arise when the signal amplitude being sampled changes by more than the step size. Dolby Laboratories Inc. (2) have devised an adaptive delta modulation (ADM) system that has been adopted for use with the B-MAC system being used over the Australian DBS service. This uses a variable step size to overcome the overloading and variable pre-emphasis to further improve the overall S/N ratio.

At the encoder of the ADM system, a pre-emphasis section analyses the frequency spectrum of the audio signal to determine the optimum pre-emphasis characteristic. After pre-emphasis, the signal is passed to a step size section, which continually evaluates the signal slope to select the step size. The pre-emphasis and step size information is coded into two low bit rate control signals. The main audio signal is then digitised and delayed by an extra 10 mS. This allows the control signals to reach the decoder in time to process the audio signal in a complementary manner.

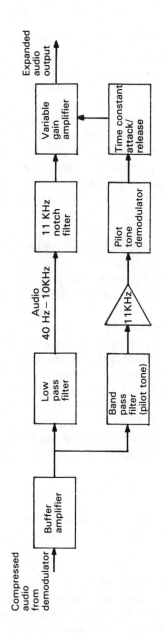

Fig. 9.6 Pilot tone controlled audio expander.

The digital data is formatted into blocks for transmission, when provision is made for synchronisation. Two types of format are provided for: one for bursty systems such as sound-in-syncs or B-MAC, and the other for continuous channels.

The basic function of the ADM decoder can be explained by Fig. 9.7. After demodulation, the signal is filtered to separate out the components. The audio data for each channel typically runs at a bit rate of 200 to 300 Kb/S and the control data at 7.8 Kb/S (half line rate). The audio data is clocked into a multiplier stage as a bi-polar signal, with the step size data acting as the muliplying constant. It is then converted into analogue format using a leaky integrator. The de-emphasis control signal functions in a similar way, but instead of being used as a gain varying element, this amplifier stage functions as a variable, single-pole frequency, de-emphasis network. The decoder, which is available as an IC – Signetics NE 5240 – is simple and relatively insensitive to component tolerances.

9.11   C-MAC (PACKET) SOUND CHANNELS

The overall frame structure for this format was described in section 7.4, to which the reader is referred. Each line period of 64 μS contains 1296 sampling points, equivalent to a sampling frequency of 20.25 MHz. The audio channels are sampled at 32 KHz and quantised into 14 bits/sample, then coded into the two's complement form. For stereo, the left and right channels are sampled simultaneously, coded separately and transmitted alternately. The sound and data bits are organised into 164 packets each of 751 bits, or a total of 123164 bits to be transmitted in 40 mS. This is equivalent to a bit rate of 3.0791 Mb/S. Taking into consideration, the corresponding bits in line 625, the actual sound/data clock rate is $198 \times 625/0.04 = 3.09375$ MHz. Allowing for the overhead of 23, header address, protection and linking bits per packet, the total sound/data capacity is:

$$728 \times 164 = 119392 \text{ bits}/40 \text{ mS} = 2.9848 \text{ Mb/S.}$$

This total capacity can be sub-divided in many ways. Depending upon the methods of coding and error protection level used, some of the possibilities include:

- 3 linear stereo channels with basic error protection
- 4 companded stereo channels with basic error protection
- 2 linear stereo channels with extended error protection
- 3 companded stereo channels with extended error protection

or the equivalent in mono or dual language channels.

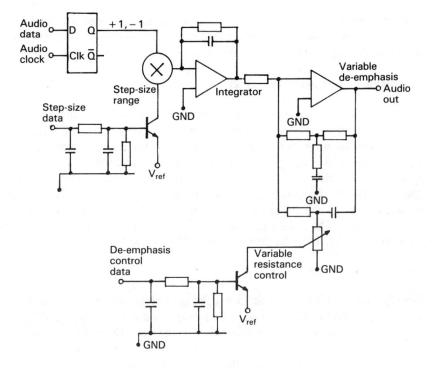

Fig. 9.7 ADM decoder.

Error Protection

Linear mode, first level: One even-parity bit is added to the 11 most significant bits of each sample.

Linear mode, second level: An extended Hamming (16.11) code is applied to the 11 most significant bits. This is capable of single-bit error correction.

Companded mode, first level: One even-parity bit is added to the first 6 most significant bits.

Companded mode, second level: An extended Hamming (11.6) code is applied to each sample. This is capable of correcting most single-bit errors.

Companding

The system used is similar to NICAM-3 (Near Instantaneous Companding Audio Multiplex, Mark-3), used for processing digital sound in the production environment. After sampling, the sound plus data burst is organised into blocks of thirty-two 14 bit samples. These are compressed to 10 bits each, using a *scaling factor* determined by the magnitude of the largest sample in the block. The scaling factor is then encoded into the parity bits for each block, to signify its degree of compression. At the receiver, this scaling factor is extracted using majority decision logic, an action that also restores the original parity. The decoded scaling factor is then used to expand all the samples in that block.

Interleaving and Energy Dispersal

The 751 bits of each packet, except those in line 625, are interleaved to minimise the effect of burst errors. After interleaving, an energy dispersal or spectrum shaping technique is applied, to randomise the data stream. This applies to all except the first 7 bits in each line and the data in lines 624 and 625. It is achieved by adding the output of a *pseudo-random binary sequence* (PRBS) generator, with a period of $2^{15} - 1 = 32767$ bits, Modulo $- 2$, to the digital bit stream, thus scrambling a block of $198 \times 623 = 123354$ bits. The PRBS runs at 20.25 MHz and is initialised every 625 lines, so that the first addition always applies to bit 8 in line 1.

Modulation/Demodulation

The digital signal is transmitted using the type of modulation known as 2–4 PSK or symmetrical DPSK, logic 1 being represented by a $+90°$ and logic 0 by a $-90°$ phase shift respectively. There are three basic ways of demodulating such signals. If the C/N ratio is high, typically greater than 16 or 18 dB, it is possible to use the vision FM discriminator to recover the audio/data signal as well. In the more common situation, either *coherent* or *differential* demodulation would be applied.

Coherent demodulators have to detect the incoming signal and compare it with a highly stable reference signal. If there is any instability here, then errors will arise. As the received signal is a form of DPSK, this problem can be avoided by using differential demodulation (3). Since the actual data is carried by the phase difference between successive intervals, this difference can easily be detected by comparing the received signal with itself after a delay of 1 bit period.

9.12 D-MAC SOUND AND DATA CHANNEL

The D-MAC specification provides for a *duo-binary* PSK signal format with the full bit rate of 20.25 Mb/s. This can be achieved from a NRZ format by using a Pre-coding technique that replaces each binary 0 with a signal transition at the bit cell centre, and a binary 1 with no transition. This new signal is then passed through a filter with a cut-off bandwidth at the half nyquist value. In this case 10.125 MHz. This generates an analogue type of signal with a small dc component. Due to filtering, adjacent transitions of opposite sense tend to cancel and leave the signal with an average zero value. A series of two or more 1's or 0's produce positive and negative peaks respectively, the duration of each peak being proportional to

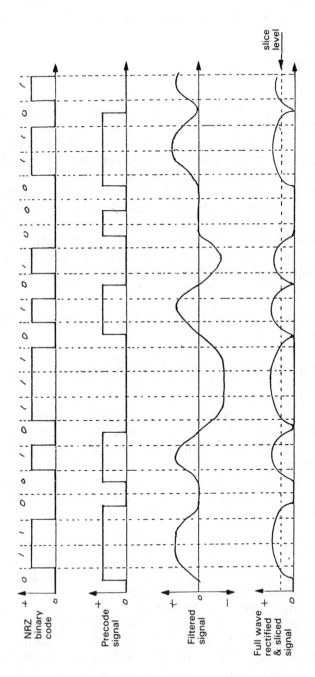

Fig. 9.8 Duobinary Signal coding and decoding.

the number of successive similar bits. The pre-coding and filtering action is displayed in Fig. 9.8. The original data signal can be recovered very simply by Full-wave rectification, followed by slicing at the half amplitude level.

D2-MAC Sound and Data Channel

At a data rate of 20.25 Mb/s, the base bandwidth of this channel would be 10.125 MHz. The time compressed vision components have a base bandwidth of 5.6 MHz × 1.5 for luminance and 2.8 MHz × 3 for chrominance, or 8.4 MHz in total. Thus, as a satellite feed to a cable system head end, the total bandwidth of more than 18.5 MHz is unacceptable to most current networks. The D2-MAC system resolves this problem by halving the data rate to 10.125 Mb/s, equivalent to a total base bandwidth of just under 13.5 MHz. This allows statellite delivered D2-MAC signals to be redistributed over most current cable networks. The reduced data capacity still leaves room for one high quality stereo channel plus a lower grade audio channel and a limited data service.

9.13   VERTICAL INTERVAL MULTIPLE CHANNEL AUDIO SYSTEM (VIMCAS)

An important feature that has to be considered when planning to add stereo audio to an established mono television network is the cost of modifying all the transmitters. An Australian organisation, IRT Ltd, have developed a *bolt-on* black box system that neatly side-steps this problem. Known by the acronym VIMCAS, this system can be used for satellite and terrestrial television and video recorders without modification to the original system.

Basically the system transmits time compressed and companded audio signals during spare line periods of the field blanking interval. Each line can support an audio base bandwidth of approximately 4.7 KHz, so that six lines can provide for a pair of stereo channels 14 KHz wide. Alternatively, multiple lines can be used for dual language or data transmissions.

Fig. 9.9 shows the general principles involved. Each audio channel signal is band limited and compressed whilst in the analogue form. The signal is then sampled and quantised and loaded into a digital memory. At the time of the appropriate video line, the memory is read out at a very much higher rate to achieve time compression. This is then converted back into analogue form and gated into the video signal. The time compressed audio signal now has a bandwidth of about 2.5 MHz, so that it is well within the capacity of the video signal channel. Reference to Fig. 9.9(b) shows that

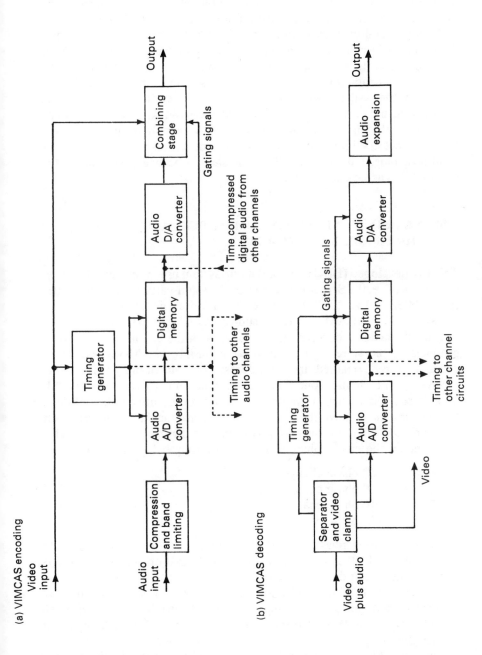

Fig. 9.9 Block diagrams for vertical interval multiple channel audio system.

decoding is achieved in a complementary manner. Any additional channels require dedicated A/D convertors and digital memories but can share the final D/A convertor, because these are used on a gated basis.

Where several contiguous lines are used for wideband audio, there is a duplication of signal at the end and beginning of successive lines. This is done to ensure that the signal at the beginning of a line, which is most likely to be corrupted with interference or distortion, can be discarded. In operation, the system has been found to be very flexible, it being possible to mix wide and narrow band signals without cross-talk. Scrambling can be provided in the digital domain of processing, or simply by alternating the line sequences. With video tape recording, the signals are not affected by head switching and wow and flutter is said to be negligible due to the method of synchronism in use.

9.14 NEAR INSTANTANEOUS COMPANDED AUDIO MULTIPLEX-728 (NICAM-728)

This UK developed (BBC and IBA) audio system uses a second sub-carrier at a level of −20 dB relative to the peak vision carrier and 6.552 MHz (6.552 MHz = 9 × 728 KHz) above the vision carrier. (As modified for NORDIC TV, NICAM-728 uses a sub-carrier of 5.85 MHz.)

The sub-carrier is differentially encoded with the digital signal for both channels of the stereo pair. The present 6 MHz (5.5 MHz) FM sound channel is still retained in the interests of compatibility with mono receivers. The digital sound carrier is quadrature (four phase) PSK modulated, where each resting carrier phase represents two bits of data, thus halving the bandwidth requirement. Because the data is differentially encoded (DQPSK), it is only the phase changes that have to be detected at the receiver, the bits to phase change relationship being as follows:

$$0\ 0 = \quad -0° \text{ phase change}$$
$$0\ 1 = \quad -90° \text{ phase change}$$
$$1\ 0 = -270° \text{ phase change}$$
$$1\ 1 = -180° \text{ phase change.}$$

Pre-emphasis/de-emphasis to CCITT recommendation J.17 is applied to the sound signal either when in the analogue format or using digital filters whilst in the digital domain.

The left- and right-hand channels are sampled simultaneously at 32 KHz, coded and quantised separately to 14 bit resolution, and transmitted alternatively at a frame rate of 728 bits per millisecond or 728 Kb/sec.

The NICAM compander processes the 14 bit samples in the manner shown in Fig. 9.10. The rule for discarding bits can be summarised as follows: the most significant bit (MSB) is retained and the four following

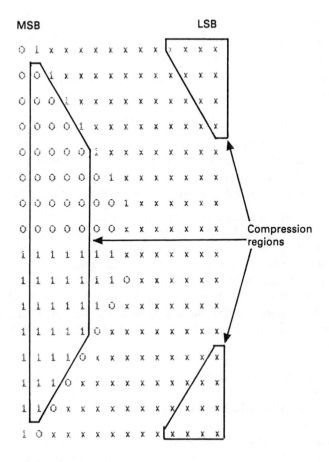

X = Don't care

Fig. 9.10 NICAM-728 companding coding scheme.

bits are dropped only if they are of the same consecutive value as the MSB. If this leaves a word longer than 10 bits, then the excess bits are dropped from the least significant bit (LSB) region.

A single even parity bit is added to check the six most significant bits in each word. The data stream is then organised into blocks of 32, 11 bit words in the 2's compliment form. A 3 bit compression scaling factor is determined from the magnitude of the largest sample in each block. This is then encoded into the parity bits for that block. At the receiver, the scale factor can be extracted using a majority decision logic circuit. At the same time this process restores the original parity bit pattern.

Two blocks of data are then interleaved in a 16 × 44 (704 bits) matrix to minimise the effects of burst errors. Adjacent bits in the original data stream are now 16 bits apart. A transmission frame multiplex is then organised in the manner of Fig. 9.11, the additional bits being used as follows:

- 8 bits are used as a frame sync word (framing word)
- 5 control bits are used to select the mode of operation, either (C_0-C_4)

    stereo signal composed of alternate channel A and B samples
    2 independent mono signals, transmitted in alternate frames
    1 mono signal plus one 352 Kb/s data channel on alternate frames
    1 × 704 Kb/s data channel, plus other concepts so far undefined.
- 11 additional data bits are entirely reserved for future developments.

Following the interleaving of the 704 sound data bits (64 × 11 bit samples) the complete frame, except the framing word, is scrambled for energy dispersal by adding Ex-Or, a PRBS of length 2^9-1, the PRBS generator being reset on receipt of the framing word.

To limit the bandwidth, the data stream is passed through a spectrum shaping filter that removes much of the harmonic content of the data pulses. This, combined with the action of a similar filter in the receiver, produces an overall response that is described as having a full or 100% cosine roll-off.

The data stream is finally divided into bit pairs to drive the DQPSK modulator of the 6.552 MHz sub-carrier.

Decoding the NICAM-728 Signal (Fig. 9.12)

The secondary sound channel sub-carrier appears at either 32.948 MHz or 6.552 MHz, depending on the method adopted for processing the receiver sound IF channel. The spectrum shaping filter forms part of the system overall pulse shaping and has an important effect on the noise immunity. The overall filtering ensures that most of the pulse energy lies below a frequency of 364 KHz (half bit rate).

The QPSK decoder recovers the data stream and the framing word detector scans this to locate the start of each frame and reset the PRBS generator. This sequence is then added Ex-Or to the data for de-scrambling. The de-interleaving process is also synchronised by the arrival of the framing word.

Error control follows standard procedures, but since this is usually buried within an IC, the process is transparent. The operating mode detector searches for the control bits C_0-C_4 to set up the data and audio stage switches automatically. The data outputs being those for the 352 or 704 Kb/s data channel options.

The NICAM expansion circuit functions in a complementary manner to the compressor, but using the scaling factor to expand the 10 bit data words into 14 bit samples.

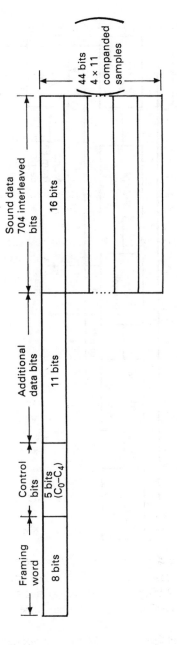

Fig. 9.11 NICAM-728 frame multiplex.

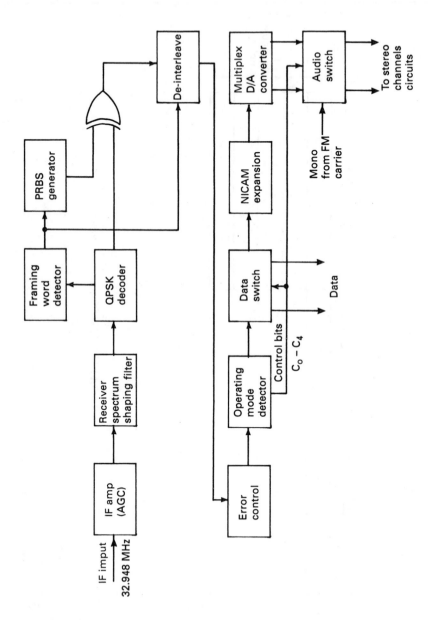

Fig. 9.12 The NICAM-278 stereo signal.

Finally, the data stream is converted back into analogue form for delivery to the audio amplifier stages. These will be designed to a very high standard, because NICAM-728 has an audio quality equivalent to that of compact disc systems.

The DQPSK decoder

This is a particularly complex stage that is available in IC form. Fig. 9.13, which represents this, is very much simplified. The two main sections are associated with the recovery of the carrier and bit rate clock. The first section relies upon a voltage controlled crystal oscillator running at 6.552 MHz and two phase detectors to regenerate the parallel bit pairs, referred to as the I and Q signals (In-phase and Quadrature). A second, similar circuit but locked to the bit rate of 728 KHz is used to synchronise and recover the data stream.

Parallel adaptive data slicers and differential logic are used to square up the data pulses and decode the DQPSK signals. The bit pairs are then converted into serial form. The practical decoder carries a third phase detector circuit driven from the Q chain. This circuit is used as an amplitude detector to generate a muting signal if the 6.552 MHz sub-carrier is absent or fails. This is then used to switch the audio system to the 6 MHz FM mono sound signal.

Apart from in Europe where NICAM-728 has been adopted as a stereo standard, the system is being installed in many areas around the world. It is also capable of modification to suit the 525 line NTSC services.

Because of the commonality the system shows with the MAC/packet systems, it should be economical to produce dual standard decoding chips. At present, at least two manufacturers have single chip decoders available (Texas Instruments and Toshiba).

9.15   WEGENER SDM 2000 DIGITAL AUDIO TRANSMISSION SYSTEM

This concept, based on Dolby ADM, is intended to provide stereo radio/audio programmes with a quality equivalent to 14 bits resolution for redistribution over a cable network. The system utilises the band space of two Wegener sub-carriers to carry two digitally encoded audio signals in a total bandwidth of about 250 KHz. A total of 512 kbit/s data stream is modulated on to a carrier using QPSK to halve the required bandwidth. This leaves space for additional channels that may be allocated as one 19.2 kbaud synchronous data, 15 remote control and 1 encryption control commands if necessary.

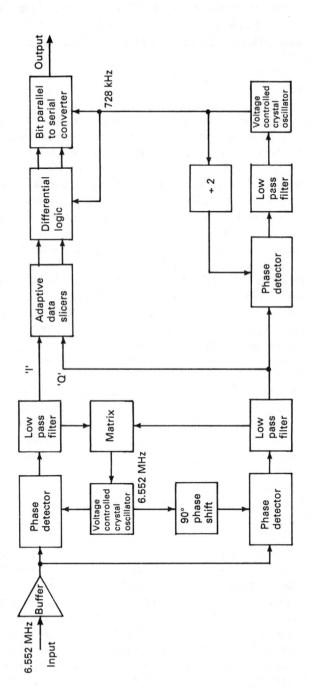

Fig. 9.13 Decoding the DQPSK signal.

9.16   SOUND RADIO BROADCASTING VIA SATELLITE (6, 7, 8)

In recent years much interest has been shown in the possibility of adding sound radio broadcasting to the satellite delivered services. The economics of introducing new, high audio quality services to the developing nations has created considerable pressure for the R & D to find an alternative to extending the present costly terrestrial networks. Again, the frequency spectrum is under continuing pressure for the expansion of such services. The extension of high quality audio derived from compact disc also generates the same demand for high-quality broadcast reception within moving vehicles. The fact that compact discs are capable of producing far higher quality than the current FM broadcast networks further adds to this pressure. The requirements also call into question the failings of FM under conditions of interference and varying propagation. Such restrictions therefore virtually rule out an analogue solution to these problems.

If an all-digital solution can be found then much of the receiver could be fabricated with a very large-scale IC (VLSI), which if produced in large world-wide quantities would result in a very low cost to the end user.

Whilst a terrestrial bound service would in many areas of the world be restricted by the local terrain, an equivalent satellite delivered service would be less affected.

The use of geostationary satellites for any service creates problems for users in high latitudes. These also require a moving vehicle to be equipped with a tracking antenna unless a low gain omnidirectional device is used. This then places a bit rate penalty on the system, automatically reducing the transmission quality. A proposed alternative is to use four (six hours spacing) synchronous satellites in a highly elliptical Molniya orbit. This would then allow for higher gain, roof-mounted antennas on moving vehicles.

A great deal of research has been carried out within Europe, North America and Japan in an attempt to create a new and world-wide acceptable system that will meet these requirements. The favoured technology for DAB (Digital Audio Broadcasting) combines *sub-band coding and masking* with *coded orthogonal frequency division multiplex* (COFDM).

One bit rate reduction technique that can be used in audio circuits relies on the *masking effect*. If two signal components are present on nearby frequencies then the ear generally responds only to the one with greatest amplitude. Therefore only the large amplitude components in any narrow bandwidth need be coded.

MASCAM (Masking-pattern Adaptive Sub-band Coding And Multiplexing) involves dividing the full audio bandwidth into a narrow sub-bands, typically using quadrature mirror filters. The spectrum of each are then continually monitored and analysed so that the audio masking effects can be predicted and provide for a bit rate reduction. Several other

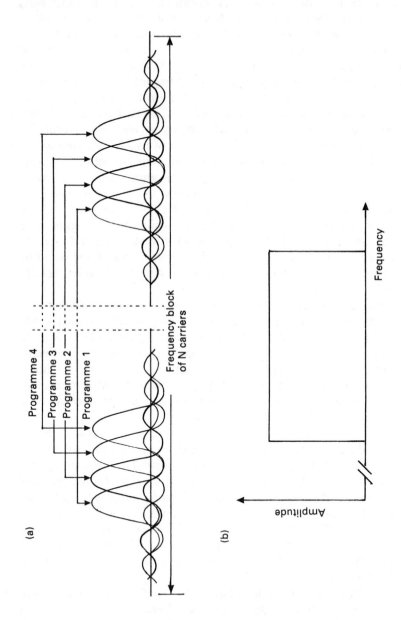

Fig. 9.14(a) Time domain of OFDM carriers. **(b)** OFDM frequency spectrum.

similar techniques have been tested. These include *ASPEC* (Adaptive SPectral Entropy Coding) and *MUSICAM* (Masking-pattern adaptive Universal Sub-band Integrated Coding And Multiplexing). The one finally chosen could well combine the best features of each.

OFDM (Orthogonal FDM)

This concept consists of generating a large number of carriers each with equal frequency spacing, with each carrier capable of being digitally modulated by one of the audio sub-band codes. The spectrum therefore behaves as a parallel transmission bus. Each modulated carrier is filtered to produce a sinx/x (sinc x) response as indicated by Fig. 9.14(a). This allows the spectra of neighbouring modulated carriers to be orthogonal and overlap as shown. When these are combined the total spectrum becomes practically flat as shown in Fig. 9.14(b) so that the channel capacity approaches the Shannon limiting value.

The allocated bandwidth is divided into N elementary frequencies and arranged to carry P programme channels. There are therefore N/P interleaved elementary carriers which carry the sub-band modulation in the manner shown in Fig. 9.14(a). Typically either four or eight phase PSK can be used to improve bit rate reduction.

Fast fourier transform (FFT) processing with convolutional coding (the C in COFDM) is employed at the modulator, with complementary FFT processing and Viterbi decoding at the demodulator. With a gross bit rate of 256 kbit/s per stereo pair, this will produce an overall bit error rate better than 10^{-3}.

Such a DAB service is expected to be allocated spectrum within the 1.5 or 2.5 GHz ranges. Developments suggest that COFDM in a 4 MHz bandwidth will support 12 stereo programmes, each with better quality and spectrum utilisation than the current analogue FM systems.

Since the COFDM spectrum has noise-like properties and can be transmitted at relatively low power, additional programmes using this technique could be transmitted within the terrestrial television taboo channels, thus expanding the capacity of this hard-pressed part of the spectrum.

REFERENCES

(1) Mountain, N. (1985) 'Satellite Transponder Operation with Video and Multiple Sub-carriers'. *Proc. 14th. International TV Symposium*. Montreux.
(2) Todd, C.C. and Gundry, K.J. (1984) 'A Digital Audio System for DBS, Cable and Terrestrial Broadcasting'. *Proc. IBC '84. 10th. International Broadcasting Convention*. Brighton. 1984.

(3) Clark, D. (1985) 'Demodulation Techniques for 20 Mb/S PSK Signals'. *IBA Report 133/85*. Independent Broadcasting Authority (UK).

(4) NICAM-728 System Specification. BBC Engineering Information Department, 1986.

(5) Texas Instruments. NICAM-728 Stereo TV Sound Decoder, 1989.

(6) Weck, Ch. et al. (1988) 'Digital Audio Broadcasting'. *Proc. IBC'88*. IEE Pub. No. 293, p. 36.

(7) Pommier, D. et al. (1988) 'Prospects for High Quality Digital Sound Broadcasting'. *Proc. IBC '88*. IEE Pub. No. 293, p. 349.

(8) Mason, A. G. et al. (1990) 'Digital Television to the Home'. *Proc. IBC '90*. IEE Pub. No. 327, p. 51.

Chapter 10

Information Security and Conditional Access

10.1 GENERAL INTRODUCTION

Due to the broadcast nature of satellite communications, some form of security becomes essential in very many applications. These include such areas as computer-to-computer data signals, as used by banks and businesses. Even the TV programme provider needs to be able to obtain revenue from his services. In many digital applications, the use of *code division multiple access* (CDMA) to a satellite transponder provides an economical way of sharing the resource. It will be recalled that CDMA involves translating the pure binary code into an alternative bit pattern. Commonly, a *pseudo-random binary sequence* (PRBS) and its inverse are used to represent logic 1 and 0 respectively. As all ground stations transmit simultaneously, only the receiver that is equipped with the correct PRBS can decode the data that is intended for it. As many bits are now required in the transmission band to represent 1 bit in the baseband, this *addressability* leads to a reduction in the actual data transfer rate.

10.2 SPREAD SPECTRUM TECHNIQUES

Theoretically, all systems where the transmitted frequency spectrum is much wider than the minimum required is a spread spectrum system (1). CDMA is therefore a digital example, and FM an analogue example. These can be shown to have an S/N ratio advantage that is gained due to the modulation/demodulation process. It was shown that FM has an S/N ratio advantage over AM of $3\beta^2 F$, where β is the deviation ratio and F the ratio of peak deviation to baseband width. In a similar way, the digital system has a processing gain proportional to the ratio R_c/R_i, where R_c and R_i are the actual transmitted code rate and the original information rate respectively.

There are two basic ways in which a PRBS can be used to generate a spread spectrum signal: the direct sequence (DS) and frequency hopping (FH) methods.

Direct Sequence Method: The original bit pattern and the PRBS are combined to produce a new high-speed binary signal that is used to modulate the RF carrier, often using a form of PSK. The transmission bandwidth is dependent upon the period of the PRBS or *chip*, as it is often known.

Frequency Hopping Method: In this system, the PRBS is used to drive a frequency synthesiser which causes the carrier to 'hop' around within the spread bandwidth. The modulator is therefore effectively a code-to-frequency convertor. The processing gain is the ratio of RF bandwidth to message bandwidth.

In both cases, the receiver has to be supplied with the correct PRBS to recover the original data. Because of the relatively slow response of the frequency synthesiser, FH systems tend to have the slower code rate – typically 200 Kb/S, as opposed to as high as 200 Mb/S for DS systems. There are also several hybrid systems in use that combine the various merits of both systems.

10.3 SCRAMBLING AND ENCRYPTION

In general, the terms, *scrambling* and *encryption* tend to be used synonymously. In this text, however, the term 'scrambling' will be taken to mean the rearrangement of the order of the original information, whilst 'encryption' will imply that the original information (often referred to as *plain* or *clear text*) has been replaced by some alternative code pattern (known as *cypher* or *encrypted text*). Scrambling alone is not considered to be secure, because a study of the signal behaviour can lead to the design of a suitable descrambler.

The encryption operation is quite simple when all the characters to be secretly transmitted are in a binary electronic form. When a second binary sequence is added to the first, using Modulo-2 arithmetic, the resulting sequence carries no obvious information. The original sequence can be revealed by performing the inverse operation at the receiver. The rules for Modulo-2 addition and subtraction can be stated as follows:

$$
\begin{array}{lll}
0+0=0 & & 0-0=0 \\
0+1=1 & \text{and} & 0-1=-1 \ (\text{ignore minus}) \\
1+0=1 & & 1-0=1 \\
1+1=0 \ (\text{ignore carry}) & & 1-1=0.
\end{array}
$$

Thus both addition and subtraction are equivalent to the logic operation of exclusive OR (EX.OR).

For example, it is required to secretly transmit the binary character 10001110 and an 8 bit key, 10101010, is chosen for encryption. The transmission/reception process then becomes:

Character to send: 10001110
Key   10101010
Sum Mod-2   00100100 (this is transmitted and received)
Key   10101010
Sum Mod-2   10001110 (the original character)

In the general case, keys are produced using PRBS generators. Such keys have several advantages, including:

- They are practically random and easy to generate and change.
- The longer the key, the more difficult becomes unauthorised decryption.

The *one-key* system just described has a significant disadvantage. The key has to be transmitted in some way, to all authorised users, before the message. This results in a time delay, but perhaps more importantly there is a risk that the key might fall into the wrong hands. Multiple key systems have been devised to reduce these risks and improve security.

In a *two-key* system, one key is made public for encryption whilst the second is kept secret and is used as a *modifier*. This is often called the *public key* system and is particularly flexible.

A very high degree of security can be achieved using a *three-key system*. Two secret keys, primary and secondary, are user-programmable and stored in a digital memory. The third non-secret key, which acts as a modifier, can be generated as a new PRBS at the start of each transmission.

Three commonly used encryption algorithms are the Diffie Hellman, the RSA Public Key Exchange System (named after the authors, Rivest, Shamir and Adelman), and the Federal Information Processing Data Encryption Standard (FIB:DES) (2 and 3).

The most important rules of any encryption system can now be stated:

(1) The number of possible keys should be very large to prevent a pirate from testing all possible keys in succession.
(2) Any fixed encryption operation should be very complex, making it impossible to deduce the operation from a few plain text/cypher text pairs.
(3) If security is to be based on secret information, then this must be created after the system is built; then if it is subsequently revealed, it will not jeopardise the entire system security.

For soft *encryption* the single key is embedded in the decoder. By comparison, the more secure *hard encryption* technique involves the use of a combination of several keys, shared between the encoder and the decoder.

The encryption/decryption process can be expressed mathematically as follows.

The encryption function E, when performed on plain text characters P, using key K, results in cypher text C. Thus the encryption process is given by:

$$E(K).P = C \qquad (10.1)$$

The decryption function D(K) is the inverse of E(K) or $E(K)^{-1}$. Therefore the decryption process is given by:

$$D(K).C = P \qquad (10.2)$$

The Diffie Hellman Algorithm (8)

This shared key system requires that each end of the communications link shall contain a limited degree of computing power to calculate the shared *secret key* S. The system contains two numbers – P, which is a prime, and X – both of which may be public.

The two communicators A and B both choose random values (a) and (b) respectively that lie between 0 and P-1.

The secure link is established as described by the following four stages:

A computes X^a mod P which is sent to B
B computes X^b mod P which is sent to A
A then computes $S = (X^b \text{ mod } P)^a \text{ mod } P = X^{ab} \text{ mod } P$
B then computes $S = (X^a \text{ mod } P)^b \text{ mod } P = X^{ab} \text{ mod } P$.

Both A and B now know the secret key S that is to be used for the encryption process.

Provided that a, b and P are large, it is extremely difficult to find the secret numbers by taking logarithms, even though X^a mod P and X^b mod P have been openly transmitted.

Rivest Shamir and Adleman (RSA) Public Key System (8)

This secure system relies upon the choice of two large prime numbers, P and Q, both chosen at random and used to calculate a modulus N number from the product PQ. Mod N can be made public without jeopardising P and Q. The encryption key (E) and the decryption key (D) are then evaluated from the formula:

ED mod $(P - 1)(Q - 1) = 1$.

The cypher text C is then obtained from the plain text M by

$C = M^E$ mod N.

The message is recovered from the received text by using

C^D mod N = M.

For example, if P = 3, Q = 11, then N = 33.

$$(P - 1)(Q - 1) = 20.$$

Thus when ED mod 20 = 1, E = 7 and D = 3 are possible values.

If M = 4, then C = 4^7 mod 33 = 16 and this is the code that is transmitted.

For decryption, 16^3 mod 33 = 4, the original code.

Recent attacks on the RSA scheme using parallel computing has not so far been successful. Increasing the keys from 100 to 150 digits would require an increase in the computing power in the order of 10^5 times. With the introduction of modern high-speed DSP (digital signal processing) ICs, RSA would appear to have a long lifetime.

Data Encryption Standard Algorithm

This algorithm, translates blocks of 64 bits of plain text into similar block sizes of cypher text, using 56 bit keys.

Each plain text block is divided into left (L) and right (R) groups, each of 32 bits, and then processed as shown in the flow chart of Fig. 10.1(a). Successive R groups are combined with successive keys using a very complex function f, which is fully described in the literature (2). Each processed R group is then added Mod-2 to the corresponding L group, the 16 bit groups being formed according to the formulas:

$$L_j = R_{j-1} \tag{10.3}$$
$$R_j = f\{R_{j-1}.K_j\} \oplus L_{j-1} \tag{10.4}$$

After processing using 16 keys, the L and R groups are recombined, but in the reverse order (R,L).

The initial 56 bit key group is divided into two 28 bit sub-groups and processed through shift register sequences. A new key is formed by combining the sub-groups after one or two left shifts. After 16 keys have been produced, each group of 28 bits is ready to repeat.

Using 56 bit keys as described, the probability of a pirate deciphering the code is as follows: P = $\frac{1}{2}^{56}$ or about 1.4×10^{-17}, very unlikely. Alternatively, $2^{56} = 7.206 \times 10^{16}$. A pirate making one attempt every nano-second, then on average it will take about 3.6×10^{16} nS, or more than 1.140 years, to decipher the message. The DES algorithm has been shown to be vulnerable to attack using massive parallel computer processing. But in many cases it must be cheaper to pay the system access charges than rent such computing power.

By using the DES algorithm repeatedly in overlapping blocks, it is possible to encypher plain text blocks that are very much longer than 64 bits.

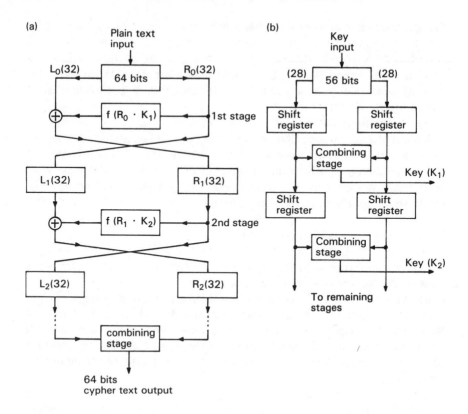

Fig. 10.1 (a) Operation of data encryption standard algorithm. **(b)** Key generator.

Although the DES algorithm is very complex, the processing can now be carried out using dedicated ICs such as the Intel 8294A (4) or the Western Digital Corp WD 2001/2 in a completely transparent manner.

Voice System Encryption/Scrambling

Within the ISDN service, it is possible to obtain security within the 64 kbit/s voice channels by using any of the above schemes. Because of the bandwidth restrictions (300–3400 Hz) of the voice grade analogue telephone network, secure communications become more difficult to achieve. However, two basic techniques are possible, using either *spectral* or *temporal* transpositions. In the former, a range of contiguous bandpass filters separate the audio signal into sub-bands. These are then either inverted or translated to new frequencies before being reassembled into a *new* analogue signal within the same frequency range but now completely unintelligible. The intended receiving terminal must be equipped with a complementary filter/processor unit.

For temporal transposition, the audio signal is converted into digital format using sampling, quantisation and delta modulation. The data stream is then divided into blocks of typically 100 msec duration which are reversed in time and then concatenated before being converted back into an encrypted analogue signal. This new signal may then be transmitted either over a telephone or radio network. A further level of security is possible by sub-dividing each block into segments which may be shuffled and then relocated within the original block period. The shuffling sequence may be controlled either from a look-up table stored in ROM (Read Only Memory), or by a PRBS generator. The synchronising of a pair of scrambled telephones is carried out using a handshaking technique to establish the scrambling sequence before the start of voice transmission.

10.4 CONDITIONAL ACCESS AND THE TELEVISION SERVICES

Scrambling in the amplitude and time domains provides the two basic methods of denying the user of a video signal its entertainment value. The most elementary method that has been used, without success, is the suppression of line or field sync pulses. This fails chiefly because modern TV receivers, with their flywheel sync/PLL type timebases, require only little modification to produce a locked picture direct from the luminance signal.

Two levels of service have to be provided for. In some countries, including the UK, a *must carry* rule applies to the cable network provider. This means that the system must distribute national programmes without restriction. If scrambling is applied to all channels, then a descrambling key has to be freely available. Therefore for premium services such as subscription or pay-per-view television, an extra level of security is needed.

It has been shown subjectively (5) that, under conditions of adjacent and co-channel interference, a scrambled picture has a 2 dB S/N ratio advantage. This is because such interference produces patterning which would apply to the scrambled picture. Descrambling then breaks up the patterning.

Video Inversion

Simple video signal inversion, which is not secure, can be successfully combined with a PRBS to provide very high security. The PAYTEL (UK) system (6) inverts alternate groups of lines in a continuously variable manner. This is effectively a three-key system, with digital data regarding the scrambling sequence being transmitted with the signal and changing with every programme. This is used in conjunction with a unique key held in each descrambling/decoder unit and the subscriber's *smart card*. The decoder key is well hidden within the decoder electronics and the user smart-card electronics can be reprogrammed on a monthly basis, the card being programmed to work only with one decoder. The card alone provides access to the must carry programmes. Typically, this credit-card-sized piece of plastic can have imbedded within it about 2 Kbytes of ROM, 1 Kbyte of EPROM, 48 bytes of RAM and a limited degree of computing ability. The electronics are thus able to keep track of the debit/credit rating of the subscriber.

Line Translation

This technique causes the line blanking period to be varied in a pseudo-random manner over a period of several frames. It requires the use of a line store, but those based on charge coupled device (CCD) technology are relatively inexpensive. To some extent, the degree of security depends upon the number of time shifts permitted within the blanking interval; systems using as few as three shifts have been shown to be insecure. The key to the PRBS for the time shifting, which is transmitted within the field blanking period, can readily be changed. The technique can be raised to the security level of a two-key system by combining the PRBS key with a *personal identity number* (PIN) that is user programmable.

The Scientific Atlanta B-MAC system, as used on the AUSSAT DBS service, uses this technique in conjunction with a three-key algorithm. Because there are no conventional sync pulses in the MAC system, it becomes very difficult for a pirate to detect where one line ends and another begins. In any case, the sync point is continually jumping around.

Line Shuffle

The lines of a normal video signal are transmitted in sequence, from 1 through to 525 or 625. If a frame or field store is available at each transmitter and receiver this order can be scrambled in a pseudo-random manner. The concept is capable of providing a very secure system, but full field or frame stores tend to be too costly for domestic television applications.

Line Segmentation

In this technique each video line is divided into segments, and it is these that are scrambled. By varying the *cut point* in a pseudo-random manner, a very high degree of security can be achieved using only two segments per line. A single 8 bit binary word can identify any one of 256 cut points. The concept is particularly economical to implement on video signals that are digital at some stage of processing. The chief constraints imposed by this method are related to the linearity and frequency response of the equipment in the transmit receive chain. Due to these imperfections, *line tilt* can arise. This is a gradual drift in the black or dc level of the signal. Under normal conditions it might pass unnoticed, but as the descrambling process changes the positions of each segment, a step appears as indicated in Fig. 10.2. Since the cut point varies in a random fashion, a visible pattern of low-frequency noise might be displayed.

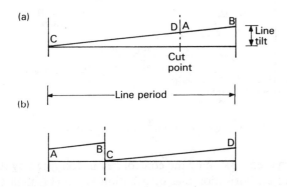

Fig. 10.2 Effect of line tilt and line segmentation scrambling **(a)** before descrambling **(b)** after descrambling.

Additionally, if such a signal is processed over a *Vestigial Side Band* (VSB) cable network, any receiver mistuning will distort the skew symmetry in the demodulator frequency response, giving rise to accentuation or attenuation of the low video frequencies. This generates overshoots or undershoots in the signal in the manner shown in Fig. 10.3, again giving rise to a noise pattern on the display.

E.B.U. Controlled Access System

The European Broadcasting Union (EBU) has devised a control standard for conditional access to the MAC/Packet system. It uses an extension of the DES three-key algorithm and its general principle is indicated in Fig. 10.4

A PRBS is defined by a *control key* that, for free access to must carry signals, is fixed and publicised, so that a receiver can automatically descramble such programmes. Where access is controlled, the control key is encrypted by an *authorisation key* which is in turn encrypted by a *distribution key*.

Decryption of the authorisation key is effected by the use of the distribution key, which may be transmitted over the air or input via a smart

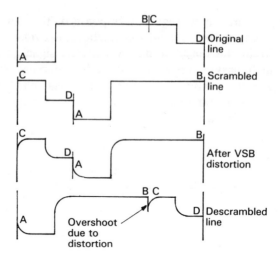

Fig. 10.3 Effect of line segmentation on VSB cable signal.

card. The encrypted version of the control key is decrypted by the use of a combination of authorisation and distribution keys. This then enables the selection of the correct PRBS to descramble the signal.

Subscriber billing can also be built into the system. This is managed via the system control computer, using reprogrammable data on the smart card, or through the distribution key.

Piracy

Probably a more insidious form of illegal operation, this involves the transmission of narrow band digital signals within the sidebands of an FM television transmission. A typical TV transponder has a bandwidth somewhat in excess of 30 MHz and at the band edges the signal contains very little power. A relatively low-power digital transmission (VSAT) with a bandwidth up to about 30 KHz is easily hidden beneath this *noise*. In an attempt to further hide these illegal transmissions by energy balancing both band edges are often employed, one used for the uplink and the other used for the downlink.

10.5 CONTROLLED ACCESS IN OPERATION

VideoCipher II (9)

This North American system developed by General Instruments Inc includes all the necessary elements for the secure delivery of high-quality vision, plus stereo audio to both CATV and DTH subscribers. It also carries all the signals necessary for system control and management, including tiering of authorised access, impulse pay per view, on-screen menus to aid subscriber choice, parental lock-out, together with text and message services.

Video security is achieved partly by removing all line and field sync pulses, inverting the video signal and positioning the colour burst at a non-standard level.

The two analogue audio channels are filtered, sampled and digitised at the standard CD rate. Each sample is added modulo-2 to a PRBS generated by the DES algorithm, forward error coded (FEC) and inter-leaved for transmission over a satellite channel. The FEC coding can detect and correct all single errors, and detect and conceal all double errors using interpolation. Together with addressing and control information, this multiplex is transmitted during the sync pulse period of each line. Since this multiplex becomes part of the 4.2 MHz video signal, no additional audio sub-carriers are needed. This effective bandwidth reduction results in an overall 2 dB improvement in video C/N ratio, relative to the unscrambled signal.

Video scrambling operates on the cut and rotate principle. The active period of each line is sampled and quantised at four times the colour sub-carrier frequency and stored in a multiple line memory. The line data is then split into segments of variable length, under the control of the DES algorithm, and the position of the segments interchanged. This data is then read out of memory and converted back into an analogue format for transmission.

To ensure system security with flexibility, a multi-level key hierarchy is employed. Each decoder has a unique public address and a number of DES keys stored within its microprocessor memory. To receive scrambled/encrypted programs for each billing period, the decoder first receives a message with the monthly key, together with service attributes (tiering, credit, etc.) This is transmitted over a control channel to the decoder with that unique key. So that only the authorised decoder can process the encrypted signal, this data is added to the decoder memory.

Every programme is encrypted with a different programme key and only those decoders equipped with the monthly key can decipher these programmes. By changing monthly keys the programme provider can automatically authorise/cancel a complete set of decoders with a single transmission.

At the programme originating centre the file of decoding and address keys is itself DES encrypted for added security. The system is so engineered that if a decoder is stolen, cloning will not allow access to programmes because pirate decoders are easily de-authorised over the air.

Eurocypher

This system, developed by European Television Encryption Ltd in cooperation with General Instruments Corp, is modelled on the VideoCipher II management and access system with enhanced security, but modified to be used with the MAC scrambling and encryption system.

An access control module within the decoder has a unique address and its memory holds a series of keys necessary to decode the signals. Other keys, which are changed on a regular basis, are transmitted over the air to provide the final level of access control.

This system is also adaptable to those PAL transmissions that are encrypted using the cut and rotate principle.

Eurocrypt

This system, developed by CCETT (Centre Commun d'Etudes de Télédiffusion et Télécommunication, France), has much in common with VideoCipher II, but is intended for use with the D2-MAC system (either cable or satellite delivery), where the primary specification provides for scrmabling on a cut and rotate basis.

Controlled access is via a smart card which is capable of being programmed over the air and which thus ultimately carries all the necessary keys for security, services and management. The final element in the key system is the control word that is transmitted within the signal multiplex, and this is changed every 10 seconds.

MAC Scrambling/Encryption

Scrambling using the double cut and rotate principle (luminance and chrominance) was included in the MAC specification because, as described earlier, this can under certain interference conditions provide an enhancement of 2 dB in the C/N ratio.

The cut point positions in each component of the video signal is controlled by a PRBS generator and an 8 bit control word is used to define 1 of 256 unique cut points. A 60 bit linear feedback shift register is used to generate a PRBS with a very long cycle time. This sequence is reset to a different starting point every 256 frames by using a control word transmitted during line 625.

The scrambling sequence is encrypted using a shared key system to provide controlled access in the manner indicated in Fig. 10.4. The data periods in the MAC packet system provides ample capacity for system management by over-the-air addressing. This allows for programme tiering, parental lock-out, decoder authorisation for specific services and pirate lock-out.

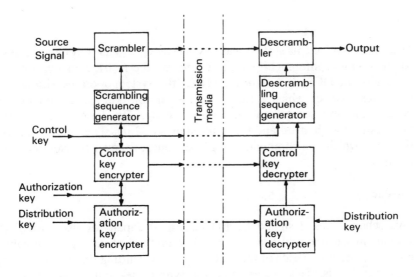

Fig. 10.4 Conditional access system for television.

The audio channel digital data stream is also encrypted in a similar way using the same PRBS generator.

Videocode and Digicrypt

These two similar systems, developed by Space Communications (Sat-Tel) Ltd, use a line shuffling technique based on a sequence of 32 or 128 lines respectively.

Each encoder and decoder has line storage for 32 or 128 lines. In addition to line shuffling, provision is made to displace randomly the line start by ± 1.5 μs. An encrypted key is transmitted in the vertical blanking interval (VBI) during lines 6, 7, 8, 9, 319, 320, 321 and 322, to instruct the decoder of the precise sequence in use.

Videocrypt/PALcrypt

This system, originally designed for PAL DTH services, also uses the cut and rotate principle and employs a code that can be changed every 10 seconds. The cut point is calculated by the system control computer and the cut and rotate performed in a 1 line memory, with the signal in the digital domain. This avoids the problems of chroma sub-carrier phase errors when each line is rejoined.

This public key system is controlled partly by data transmitted during the VBI and partly by a code held in a smart card which has to be changed periodically.

The audio channel is not generally encrypted.

B-CRYPT

The algorithm for this system was devised by GPT Video Systems Ltd for use with videoconferencing and to overcome the restrictions imposed on the use of the DES algorithm. Like DES, this system operates with two 56 bit codes, one of which forms a crypto-variable key calculated by the system and the second an initialisation code. The latter is a random number that is changed and transmitted every 32 ms. The overall encryption key, which is a combination of both codes, is never transmitted over the link.

Sat-Pac

This scheme was designed by Matsushita of Japan for use with PAL systems in Europe and Scandinavia. The method of encryption involves inverting alternate fields of the video signal and shifting the dc level of the blanking signals. This ensures that the sync pulses and colour burst are buried within the video signal level.

The composite blanking signals are then transmitted on an FM 7.56 MHz sub-carrier. A second sub-carrier at 7.02 MHz is used to carry data to enable/disable viewers decoders according to their subscription levels.

REFERENCES

(1) Dixon, R.C. (1976) *Spread Spectrum Systems*. New York: John Wiley & Sons.
(2) 'DES Algorithm'. (1977) *FIBS: PUB 46*. Washington, DC: National Bureau of Standards, Data Encryption Standards.
(3) McArdle, B.P.(1986) 'Using the Data Encryption Standard'. *Electronics and Wireless World*. London: Business Press International Ltd.
(4) 8294A Data Encryption Unit, Intel International Corp. Ltd, Swindon, UK. Applications note.
(5) Edwardson, S.M. (1984) 'Scrambling & Encryption for Direct Broadcasting by Satellite'. *Proc. IBC '84. International Broadcasting Convention* Brighton, UK.
(6) Dickson, M.G., Paytel Ltd, Reigate, Surrey, UK. 'How the PAYTEL System Works'. Private communication to author.
(7) Mason, A.G. (1984) 'Proposal for a DBS Over-Air Addressed Conditional Access System'. *IBA Experimental Report 132/84*. Independent Broadcasting Authority, UK.
(8) Denning, D. (1983) *Cryptography and Data Security*. Addison Wesley.
(9) Katznelson, R.D. VideoCipher Division, General Instruments Corp, San Diego, USA. Private communications to author.

Chapter 11

Installation and Servicing

11.1 INITIAL SURVEY

The installation of a small dish satellite-receiving ground station can produce problems that involve several different classes of regulations. These may range from structural engineering, town planning/use of land regulations, to safe working practice, both on site and during transportation. A very useful code of practice that relates to these and many other associated problems is to be found in the literature (1).

The site survey should include a check to ascertain a clear view of the equatorial orbit, in particular in the direction of the satellites of interest. It is not only important to ensure that there are no obstructions such as buildings, trees, etc, but also that there are no plans for such construction in the future. Nearby roads that are used by high-sided vehicles can also be a problem. Generally, because of the great path length between receiver and satellite, there is no advantage to be gained from a roof mounting unless it is necessary to clear some obstruction. In such a case it will be necessary to consult structural engineers to ensure that the roof is strong enough to support the load. A ground site will need to be prepared that is capable of supporting the antenna structure and this may lead to considerations of site access.

Having located a suitable site, it is important to check for any local sources of terrestrial microwave interference. This is particularly so for C Band installations. This can be done by temporarily inter-connecting the system, using the LNB without its reflector antenna, and scanning the surrounding area for signs of microwave transmissions. On a television receiving system, a no-interference situation will be indicated by a simple random noise display. Interference will produce a display with horizontal banding as the receiver is tuned across the frequency band. If interference is found, it may be possible to resite the antenna, making use of the shielding effects of a building or even a mound of earth. It is much more economical to spend time at this stage than to become involved in an expensive filtering operation when the antenna is installed.

Interference and poor C/N ratios are much more likely to be encountered at the higher-latitude sites, as the antenna bore-sight is *looking* along a noisy earth. Limiting locations are in the order of 75° of latitude and longitude difference between the orbital and earth station positions.

11.2 MECHANICAL CONSIDERATIONS

The extent of the site preparations needed depends upon the particular application. CATV, SMATV and multi-service digital systems demand an antenna of 2 to 4 metres in diameter, and one per satellite; small dish business systems and domestic TVRO installations will generally work effectively with antennas of less than 1.2 metres diameter. The operational application will largely define the type of mount to be used. For the larger fixed antenna, the 'A' frame type of mount is most suitable. However, its concrete base positioning must reflect the limited degree of azimuth adjustment available once construction is complete. For smaller installations, the azimuth/elevation mount is the simplest to set up, but if signals from more than one satellite are required then two servo steering motors will be needed. In such cases, a polar mount, which requires only one drive system, is more flexible but more difficult to set up. The king-post/pedestal component of these mounts must not only be very rigidly fixed but must also be very accurately vertical.

There will be considerable stresses set up on the mounting, due to the wind loading and aerofoil effects on the dish. An adequate safety factor must be allowed for. This should take into consideration the following problems.

(1) Metal fatigue can arise from excessive vibration and flexing.
(2) Aluminium can be subject to acid attack if exposed to a sulphurous atmosphere.
(3) Electrolytic corrosion results from the contacts between dissimilar metals.
(4) Galvanised steel will corrode if the zinc coating is damaged.

The whole structure should be adequately protected by paint and at the same time effectively earthed.

A further important electrical consideration at this stage is the distance between the head end and the main receiver. If this exceeds about 100 metres, then almost certainly a line amplifier will need to be provided at the LNB IF output.

11.3 ANTENNA/LNB COMBINATION

Just as the mechanical considerations are crucial to the system's effective operation, so is the actual choice of antenna/LNB combination. This is effectively the first stages of the system and so greatly affects the overall signal performance. Mistakes made at this stage will result in either an over-engineered system or one that gives a poor performance. In either case, this results in a dissatisfied client and/or expensive reworking. A theoretical evaluation is critical if this is to be avoided.

(1) Consult the published footprints for the satellites of interest to determine the available signal levels.
(2) Calculate the azimuth/elevation angles and the range, as set out in Appendix A3.3, or consult the graphs of Figs. 11.1 and 11.2. In particular, calculate the worst case path length or range.
(3) For the worst case range r, calculate the free space attenuation A for the frequency range involved, using:

$$A = \{1/(4\pi r)\}^2, \text{ or}$$
$$A = 20 \log \{1/(4\pi r)\} \text{ dB} \tag{11.1}$$

(4) Make allowances for additional losses due to pointing errors and rain. Typically, 2 to 3 dB is adequate, except for the case where the same rain storm can affect both up and down links simultaneously, when an allowance of 4 to 5 dB should be made.
(5) Using the receiver's required C/N ratio as a basis, calculate the down-link power budget to evaluate the G/T ratio. The antenna/LNB combination's gain and noise figure/noise temperature will dominate the system's noise performance, so that a C/N ratio 2 dB above the demodulator threshold can be aimed for.
(6) The G/T ratio can be provided in many ways, ranging from a large antenna with an inexpensive LNB to a small antenna with a very low noise LNB. The final compromise will be based on what is technologically and economically possible.
(7) The antenna diameter can now be calculated, or obtained from Table 11.1.

Table 11.1

Diameter (metres)	Gain (dBi)		Beamwidth (degrees)
	P/F fed $\eta = 55\%$	O/S fed $\eta = 70\%$	
0.5	32.6	33.7	3.5
1.0	38.6	39.7	1.7
1.5	42.2	43.2	1.2
2.0	44.7	45.7	0.9
2.5	46.6	47.6	0.7
3.0	48.2	49.2	0.6
3.5	49.5	50.6	0.5
4.0	50.7	51.7	0.45

P/F = prime focus feed; O/S = offset feed

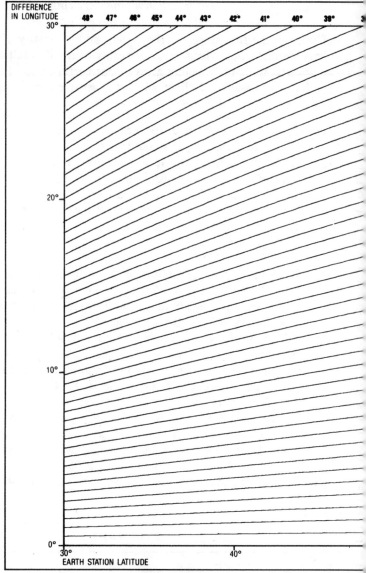

48° 47° 46° 45° 44° 43° 42° 41° 40° 39° 38°

30°

20°

10°

0°
30°   40°
EARTH STATION LATITUDE

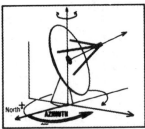

AZIMUTH ANGLE

The azimuth angle is the angle by which the antenna, pointing at the horizon, must be rotated about its vertical axis. Rotation must be in a clockwise direction, starting from geographical North, to bring the radiation axis of the antenna into the vertical plane containing the satellite direction. The azimuth angle is between 0° and 360°. The values given on the chart are

added to or subtracted from 1... depending on whether the sta... lies to the East or to the We... the satellite meridian.

Fig. 11.1 Azimuth angle (courtesy of European Telecommunications Satellite Organisation 'EUTELSAT').

| | 36° | 35° | 34° | 33° | 32° |
| | | | | | | 31° |

31°
30°
29°
28°
27°
26°
25°
24°
23°
22°
21°
20°
19°
18°
17°
16°
15°
14°
13°
12°
11°
10°
9°
8°
7°
6°
5°
4°
3°
2°
1°

60°   70°

<table>
<tr><td colspan="2" align="center">EXAMPLE 1</td></tr>
<tr><td>dinates of the site
   Latitude
   Longitude</td><td>42° North
14° East</td></tr>
<tr><td>itude of satellite</td><td>7° East</td></tr>
<tr><td>rence in longitude</td><td>7°</td></tr>
<tr><td>ing given on the chart
7 , 42)</td><td>10.4°</td></tr>
<tr><td>ion relative
itellite meridian</td><td>East
↓</td></tr>
<tr><td>uth angle</td><td>180° + 10.4°
= 190.4°</td></tr>
</table>

<table>
<tr><td colspan="2" align="center">EXAMPLE 2</td></tr>
<tr><td>Coordinates of the site
   Latitude
   Longitude</td><td>52° North
3° West</td></tr>
<tr><td>Longitude of satellite</td><td>13° East</td></tr>
<tr><td>Difference in longitude</td><td>16°</td></tr>
<tr><td>Reading given on the chart
for (16   52)</td><td>20.0°</td></tr>
<tr><td>Position relative
to satellite meridian</td><td>West
↓</td></tr>
<tr><td>Azimuth angle</td><td>180°   20.0°
= 160.0°</td></tr>
</table>

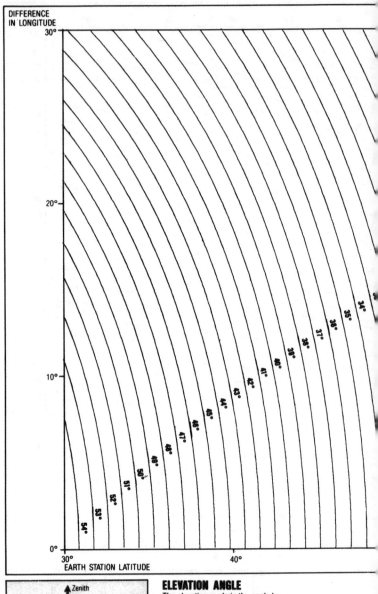

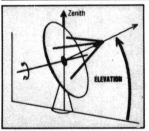

ELEVATION ANGLE
The elevation angle is the angle by which the radiation axis of the antenna must be rotated vertically, from its horizontal position, to align it with the direction of the satellite. The elevation angle is between 0° and 90°

Fig. 11.2 Elevation angle (courtesy of European Telecommunications Satellite Organisation 'EUTELSAT').

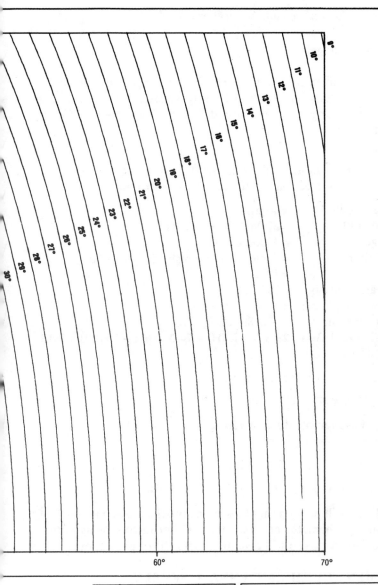

9°
10°
11°
12°
13°
14°
15°
16°
17°
18°
19°
20°
21°
22°
23°
24°
25°
26°
27°
28°
29°
30°

60°   70°

EXAMPLE 1	
Coordinates of the site	
Latitude	40° North
Longitude	33° East
Longitude of satellite	7° East
Difference in longitude	26°
Reading given on the chart for (26, 40)	36.6°
Elevation angle	= 36.6°

EXAMPLE 2	
Coordinates of the site	
Latitude	48° North
Longitude	7° East
Longitude of satellite	8° West
Difference in longitude	15°
Reading given on the chart for (15; 48)	33.0°
Elevation angle	= 33.0°

Down-Link Power Budgets

The power budgets given in chapter 6 can be rearranged to analyse the antenna/LNB characteristics, Table 6.1 being suitable for dealing with digital systems. Rearrangement of the budget for a TV system yields the following relationship:

Received C/N ratio =
G/T + EIRP — Boltzman's constant — noise bandwidth
— free space attenuation — other losses

Assuming a satelite EIRP = 46 dBW
Bandwidth	= 27 MHz = 74.3 dBHz
Range	= 39000 km
Required C/N ratio	= demodulator threshold (8dB) + 2dB
	= 10 dB

Free space attenuation A = 205 dB
Rain and pointing losses = 3 dB
Boltzman's constant = − 228.6 dB/K/Hz
10 dB = G/T + 46 dBW + 228.6 dB/K/Hz − 74.3 dBHz − 205 dB − 3 dB
G/T = 17.7 dB/K

If the chosen LNB has a noise factor of:

$$1.5 \text{ dB} = 10 \log (1 + T/T_0)$$

equivalent to a noise temperature of 120 K.
Assume an antenna noise input of 60 K, then:

Total noise = 180 K = 22.6 dBK

Thus required antenna gain = 22.6 + 17.7 = 40.3 dBi.

From Table 11.1, it will be seen that this can be provided by an antenna of less than 1.5 metres diameter of either type. Other factors being equal, however, the offset feed type will give an extra 1 dB of protection.

11.4 ALIGNMENT OF THE ANTENNA ASSEMBLY

Having derived suitable parameters for the various elements of the system, the antenna can then be installed in accordance with the code of practice. The next stage involves the four adjustments of azimuth or bearing, elevation/declination, focus of the LNB feed, and the polarisation skew, where linear-polarised signals are of concern. The data obtained from Figs 11.1 and 11.2 will be found useful.

The initial pointing accuracy in the first two stages depends upon obtaining the values of latitude and longitude for the antenna site to within 0.1°. The calculated values of the azimuth/elevation angles thus obtained can be finally rounded to the nearest 1/4° or 1/2°, depending on the beamwidth of the antenna. Because the direction of the azimuth bearing depends upon the fixed stars, it is necessary to locate the true north bearing. Magnetic north varies with time and place. For instance, in the south-east of England during 1990, the magnetic north is about 4.5° west of true north, and this could decrease by about 1/2° during the following three years. The magnetic north pole is located close to longitude 101°W, latitude 76°N, near to Bathurst Island, off the northern coast of Canada. As a result, the *magnetic declination*, or deviation between true and magnetic north bearings, varies world-wide, the error ranging from zero to more than 40° east or west of true north. The deviation for any particular location can be obtained from aeronautical navigation charts.

Even if due allowance is made for this, the magnetic compass can be grossly misleading near to iron structures. Errors of more than 20° have been noted near to steel-framed buildings, and correspondingly smaller errors close to antenna mountings. In the UK, Ordnance Survey maps are available (similar are available in most other countries) and these can often be used to derive a bearing from some prominent local landmark. However, because of the projection used for such maps, the direction of grid north can vary from true north. Figure 11.3 shows the relationship between these northerly directions, with typical differences for the south-east of England.

Solar Transit

Alternatively, use may be made of the sun's *meridian passage* or *transit* – the time when the sun is due south. Nominally, the sun is due south of the Greenwich meridian at 12.00 hours Greenwich Mean Time (GMT), or Universal Time (UT). Here due allowance needs to be taken for any local summer daylight saving arrangements. In the UK, for example, British Summer Time (BST) = GMT + 1 hour. Since the transit occurs approximately every 24 hours, the sun traverses the skies at the rate of 360°/24 = 15° per hour, or 1° per 4 minutes.

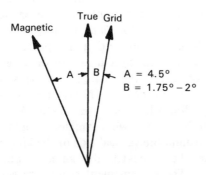

Fig. 11.3 Relationship between various definitions of 'north'.

This simple relationship does not take into consideration the cyclic variations in the solar transit. This can be as much as ± 15 minutes from GMT, depending on the time of year. Thus an error of $\pm 3.75°$ can easily be made. In astronomy, this effect is described by the *equation of time*. The due south time, or solar transit time, is tabulated in such as Reed's or Whitaker's nautical almanacs. These show the sun's bearing from Greenwich at 12.00 hours GMT for each day of the year, and this is referred to as the *Greenwich hour angle* (GHA). From this data, the due south time can easily be calculated. For example, if the table for a particular day gives a large number such as 358° for GHA, this means that the sun is 2° east of the meridian at 12.00 hours. The local noon, and hence the due south direction at this longitude, occurs 8 minutes early, at 11.52 a.m. A small number, such as 3°, indicates that the local noon is later by 12 minutes, so that the sun will be due south at 12.12 p.m. The difficult problem of finding an accurate direction thus becomes the simpler one of finding an accurate time. The solar transit time can also be used to calculate the time that the sun will occupy the same azimuth bearing as a wanted satellite.

Having calculated the local noon time, the difference between the local longitude and the satellite's azimuth has to be traversed by the sun at 1° per 4 minutes. Assuming that the satellite azimuth is 20° west of the ground station, the sun will be on the same bearing after $20 \times 4 = 80$ minutes after the local noon time.

Such information can be useful for the local calibration of a magnetic compass.

The elevation angle can be set using an *inclinometer* set against the vertical edge of the antenna aperture plane, remembering to take into consideration the typical 28° difference between the bore sight and aperture plane when installing an offset fed antenna.

Polar Mounts

In relation to polar mounts, some alternative terminology may be encountered. The elevation due south and the azimuth adjustment are sometimes known as the *axis inclination* and *east/west steering angle* respectively. The information for setting up a polar mount is derived from Appendix A3. The total declination angle, for the given latitude (ϕ) can be calculated ($\alpha + \beta$) from the graphs of Fig. A3.5. This angle should then be set on the declination adjustment. Next, the elevation angle can be calculated from equation A3.11, $90° - (\alpha + \beta + \phi)$. This allows the angle of the aperture plane to be set. The whole antenna should then be directed due south, by rotation around the vertical axis of the king post/pedestal. Once this adjustment has been locked, the antenna will swing around its declination bearings. These adjustment points are clearly shown in Fig. 11.4.

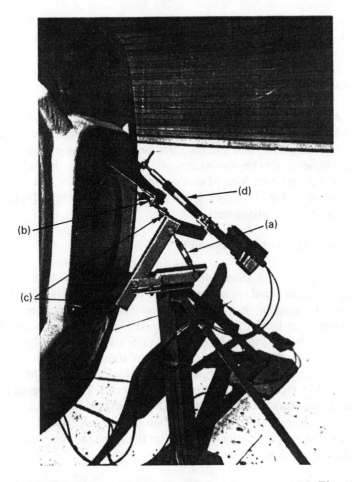

Fig. 11.4 Adjustment points for modified polar mount. (**a**) Elevation angle adjustment. (**b**) Declination angle adjustment. (**c**) Pivot axis. (**d**) Jackscrew and motor for azimuth adjustment.

Focus Adjustments

The focal point of a parabolic reflector can be calculated from the dimension of the dish, using the formula:

$$F = r^2/4d$$

where

F = distance of focal point from centre of dish
r = radius of antenna aperture plane
d = depth of dish

This gives the approximate practical position at which the LNB feed horn, or the sub-reflector, needs to be initially placed before final adjustment for maximum antenna gain.

Polarisation Skew

In the case of reception of linear polarised signals, the waveguide pick-up probe needs to be positioned parallel to the plane of the signal electric field. Only if the receiving antenna and the satellite share the same longitude will the polarisation angle be the same as that at which the signal was transmitted. For receivers located east or west of the satellite, the angle will rotate as shown by Fig. 11.5, which can be used to obtain initial setting of the polarisation skew adjustment.

Fine Tuning of the Antenna Assembly

The final tuning of all of these adjustments should be made with reference to the received signal level. A signal strength meter driven from the receiver AGC circuit is suitable, but not very convenient. The receiver may be situated a long way from the antenna. The AFC system can produce some misleading effects, so should be switched OFF during these adjustments. However, the FM capture effect can lead to similar problems. A small readjustment can suddenly produce complete loss of signals that do not return when the original adjustment position is reset. This is caused by the receiver demodulator running out of the capture effect *hold range*. For the peaking of the polarisation skew, it is often best to tune up on the wrong polarisation where it is usually easier to see the minimum response, rather than the maximum level on the right polarisation.

Spectrum Analysers

Normally, the oscilloscope provides a window in the time domain, but when used with a spectrum analyser it can provide a window in the frequency domain. This makes it particularly useful for setting the antenna adjustments. However, until recently this expensive instrument was chiefly restricted to laboratory use. The AVCOM Inc. (Richmond, Va. USA) PSA-35 Spectrum Analyser utilises the system LNB for most of its microwave signal processing, and this considerably reduces the cost. It is battery operated and provides power for the LNB, which makes it readily portable and suitable for on-site use.

Satellite Tracking Meters

There are now many inexpensive instruments available, designed specifically as aids to antenna alignment and adjustment. They are basically portable receivers designed to operate with IF inputs in the range 950 to 1750 MHz, sometimes in two bands and giving various indications of received signal strength. They are commonly operated from rechargeable batteries and suitable for one-person operation. The main receiver section is often broadband tuned to remove this variable from the exercise.

In operation, the antenna is first approximately positioned according to the azimuth and elevation calculations and the meter plugged into the LNB. This action provides power for the LNB and a signal input to the instrument at the same time. At this stage the instrument usually gives some audible indication of the presence of a satellite signal, an audio channel being helpful in identifying a particular transponder transmission. As the antenna adjustments approach optimum the pitch or loudness of the audio tone may increase. Additional indication may be given on bar graph or analogue type meters. The receiver section also often has an automatic gain reduction facility that adjusts sensitivity as the received signal level increases.

Weather-proofing

Before leaving the antenna site, ensure that all mounting assembly nuts and bolts are securely locked. Weather-proof all external electrical connections, including the co-axial cable connectors and power leads to the polarisation selector device. The exposed lead screws and pivot bearings of the steering mechanism should be adequately protected with grease.

Polar Mount Steering Drives

There are various ways of applying remote control to the polar mount steering mechanism, and these have a bearing on servicing. The basic drive assembly consists of a geared, fractional horsepower dc motor, driving a jack lead screw, to swing the antenna around its mounting axis. See Fig. 11.4. The simplest arrangement involves using a variable resistor transducer attached to motor drive, so that its resistance value is proportional to the antenna position. The remote control unit also contains a similar valued variable resistor which is attached to a calibrated angular scale. The two resistors form part of a bridge circuit which, when balanced, indicates that the antenna points in the direction indicated by the remote control. The balance condition is used to halt the motor drive. An alternative version that employs a similar transducer uses the voltage developed across it to indicate the antenna position. This voltage is converted into a digital value that can be compared with a preset value stored in the control system memory to drive the steering system.

Two further variants use a pulse-counting technique to control antenna position. Permanent magnets attached to the motor drive gears either energise reed relays or Hall effect devices to generate a pulse train. The

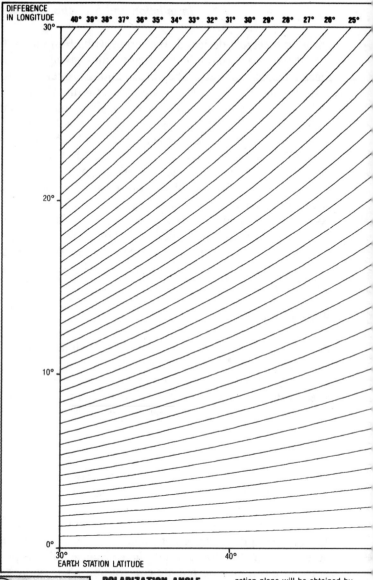

40° 39° 38° 37° 36° 35° 34° 33° 32° 31° 30° 29° 28° 27° 26° 25°

30°

20°

10°

0°

30°   40°

EARTH STATION LATITUDE

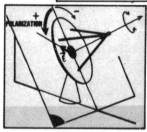

POLARIZATION ANGLE

This chart allows us to determine the angle between the polarization plane of a linear polarized wave emitted by the satellite and the vertical plane containing the antenna axis (once the antenna is pointing at the satellite).

The values given on this chart correspond to a situation where the polarization plane is perpendicular to the orbital plane. In this case, the position of the polari-zation plane will be obtained by rotating from the vertical plane in a clockwise direction (for an observer positioned behind the antenna and looking towards the satellite) if the station lies to the East of the satellite meridian, and in an anti-clockwise direction if the station lies to the West of that meridian.

If the polarization plane of the emitted wave is not perpendicu-lar to the orbital plane, a further

Fig. 11.5 Polarization angle (courtesy of European Telecommunications Satellite Organisation 'EUTELSAT').

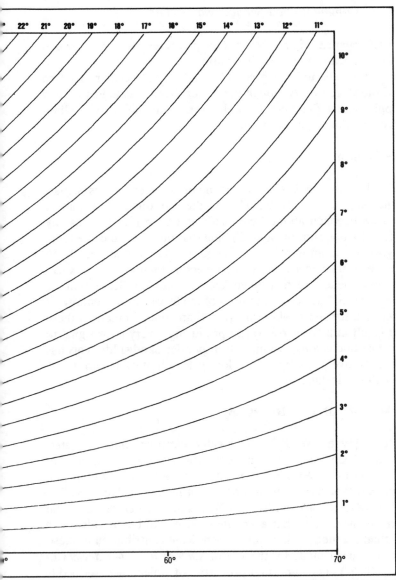

10°

9°

8°

7°

6°

5°

4°

3°

2°

1°

60°   70°

rotation will be necessary from the position thus obtained. For the EUTELSAT 1 satellites this will be 3.5° or 93.5° in a clockwise direction, depending on the polarization used (Polarization Y or Polarization X, respectively). For the TELECOM 1 satellites it will be 22° in an anti-clockwise direction.

EXAMPLE 1	
Coordinates of the site	
Latitude	50° North
Longitude	8° East
Longitude of satellite	13° East
Difference in longitude	5°
Reading given on the chart for (5: 50) 4.2°	
Position relative to satellite meridian	West
	↓
Polarization angle	– 4.2°
For the EUTELSAT I satellites the polarizations will therefore be as follows	
Pol X 93.5° – 4 2° = 89.3°	
Pol Y 3 5° – 4 2° = –0.7°	

EXAMPLE 2	
Coordinates of the site	
Latitude	50° North
Longitude	6° East
Longitude of satellite	8° West
Difference in longitude	14°
Reading given on the chart for (14: 59) 8.4°	
Position relative to satellite meridian	East
	↓
Polarization angle	+ 8.4°
For the TELECOM 1 satellite the polarization will be:	
– 22 0° + 8.4° = – 13.6°	

number of pulses generated, in respect to some preset reference value stored in memory, then indicates the antenna position.

It is important that a periodic inspection be made of the weather-proofing of this part of the structure. Although the motor drive mechanism usually contains a slipping clutch that operates in the event of the failure of the limit switches, mechanical damage can occur.

Coaxial Cable Problems

The coaxial cable between LNB and the main receiver carries signals typically in the range 950 to 1750 MHz. At these frequencies a quarter wavelength varies between about 4 cm (1.6″) and 8 cm (3.2″). Thus any sharp bends in the cable run, or tight clipping that distorts the cable cross section, can give rise to an impedance change, leading to a mismatch. In addition, the use of incorrect cable end connectors or poor fitments can further create mismatches. Such problems create multiple reflections along a cable run, leading to destructive additions of the forward wave energy. It has been noted that for short cable runs (less than about 10 metres) these problems can result in a Ku band system producing a very poor signal to noise ratio on just one channel within a transponder bandwidth. This has almost always been overcome by using a longer length of cable and taking greater care with the installation.

11.5 SURFACE MOUNTED DEVICES

In the decade following WARC '77, greater improvements in system performance have been achieved than were anticipated at that time. The fixed satellite services (FSS) were thought then to be only capable of supporting services that used antennas of 3 to 4 metres in diameter. The technological advances, particularly in LNBs and receiver demodulator threshold performance, show that adequate services can be provided with antennas less than 1.2 metres diameter. A significant contribution to these improvements has been made by the use of *surface mounted devices* or *components* (SMD/C). The important properties of which are described in section 4.8. Figure 11.6, provided by courtesy of Mullard (UK) Ltd, shows the degree of space saving that can be achieved with these devices. Allied to the use of SMDs are the changes in soldering technology that have occurred over the past two decades. What was at one time seen as an acquired practical skill is now seen as the application of a section of chemical science, a feature that has also made a significant contribution to the improvements in system performance.

The use of smaller circuit boards also leads to a further small improvement in RF performance. The printed circuit boards (PCBs) or substrates used with SMDs normally carry no through holes for

Fig. 11.6 Surface-mounted devices (courtesy of Philips Components Ltd, formerly Mullard Ltd).

conventional components; the component lead outs are soldered directly to pads or lands provided on the metallic print. This feature has led to the development of new soldering techniques which impart further advantages. In manufacture, surface mounted technology readily adapts to automated assembly, and the soldering methods used lead to improved connection reliability, both of which mean a reduction of costs. The particular technique employed in manufacture has a bearing on the way in which SMDs can be handled during servicing.

Wave soldering: The components are attached to the solder resist on each PCB, using an ultra-violet light or heat-curing adhesive. The boards are then passed, inverted, over a wave soldering bath with the adhesive holding the components in place, whilst each joint is soldered.

Reflow soldering: A solder paste or cream is applied to each pad on the circuit board through a silk screen, and components are accurately positioned and held in place by the viscosity of the cream. The boards are then passed through a melting furnace or over a hot plate, to reflow the solder and make each connection.

Vapour Phase Reflow Soldering: This more controlled way (2) of operating the reflow process uses the latent heat of vapourisation to melt the solder cream. The boards to be soldered are immersed in an inert vapour from a saturated solution of boiling *fluorocarbon liquid*, used as the heating medium. Heat is distributed quickly and evenly as the vapour condenses on the cooler board and components. The fact that the soldering temperature cannot exceed the boiling point of the liquid (215°C) is an important safety factor.

Solders and the Effects of Component Heating

Electronic components experience distress at all elevated temperatures (3). For example, in wave soldering, most baths have an absolute limit of both temperature and time – normally, 260°C for no more than 4 seconds. As the damaging effects of heat are cumulative, manufacturing and servicing temperatures have to be kept to a minimum. The use of a low melting point solder is therefore crucial. The common 60/40 tin/lead solder (MP = 188°C) is not suitable for SMD use. The lowest melting point tin/lead alloy (eutectic alloy) solder has a melting point of 183°C, and this also is not suitable. Components often have silver or gold-plated lead-outs to minimise contact resistance. Tin/lead solder alloys cause *silver leaching*; that is, over a period of time the solder absorbs silver from the component and eventually causes a high-resistance joint. This can be avoided by using a silver-loaded solder alloy of 62% tin, 35.7% lead, 2% silver and 0.3% antimony. Such a solder has a melting point of 179°C. Tin also tends to absorb gold with a similar effect, and this is aggravated by a higher soldering temperature. This latter alloy is thus particularly suitable.

The fluxes used as anti-oxidants are also important. An effective flux improves the solderability of the components, the rate of solder flow and hence the speed at which an effective joint can be made. The flux used for SMD circuits should either be a natural organic resin compound, or one of the newer equivalent synthetic chemical types.

Servicing SMD Circuits

For the larger centralised service department, a soldering rework station might well be cost effective. This might include a small portable vapour phase soldering unit, such as the Multicore Vaporette (3), which is particularly suited for small batch work as commonly found in such establishments. For the smaller service department, much ingenuity might be needed when dealing with SMD circuits.

The method used to remove suspect components may vary with the number of leads per device. For a device with only two or three leads, a fine soldering iron, in conjunction with a *solder sucker* or *solder braid* can be successful. For multi-pin ICs, two methods are popular. One involves directly heating the soldered connections with a carefully temperature-controlled hot-air blast. The other method uses an electrically heated collett or special extension to an electrical wire stripper, to heat all pins simultaneously. When the solder flows, the component can be lifted away. In a simpler way, all the leads can be severed using a pair of cutters with strong, fine points. The tabs can then be removed separately. This is not quite so disastrous as it appears. Any suspect component that is removed, and subsequently found not to be faulty, should *not* be reused: the additional two heating cycles are very likely to lead to premature failure. As the desoldering of even a few leads can be difficult, it is important that the circuit board should be firmly held in a suitable clamp.

Even after all the solder has been removed, a problem of removing the component may still exist if an adhesive was used in manufacture. Although care should be exercised when prising the component off the board, damage to the printed tracks is unlikely. The adhesive should have been applied only to the solder resist. After component removal, boards should be examined under a magnifier to check for damage to the print.

Before commencing to fit the new components, the solder pads should be lightly re-tinned, using the appropriate solder and flux. Each component will need to be precisely positioned and firmly held in place whilst the first joints are made. For multi-leaded components, secure two diagonally opposite leads first. The soldering iron used should not exceed 40 watts rating and should not be applied to a joint for more than 3 seconds. The heat from the iron should be applied to the component via the molten solder and not direct as indicated, in Fig. 11.7(b). The final joint should have a smooth 45° angled fillet. Figure 11.7 also shows three incorrectly made joints: (a) has used too little solder, and this produces a high-resistance connection; (c) has been made with the application of too little heat, which will also lead to a high-resistance connection; in (d), too much solder has been applied and this has probably resulted in excessive heating of the component.

Some SMDs are too small to value-code in the conventional way. These should therefore be stored in their packaging until actually needed, to avoid mixing components that are difficult to identify.

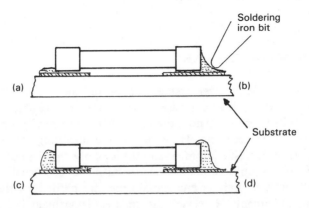

Fig. 11.7 Soldering of 'chip' type SMDs.

11.6   CONTROL OF STATIC IN THE SERVICING ENVIRONMENT

MOSFETs, CMOS and Group iii/v (GaAs, etc.) semiconductor devices can easily be damaged by the discharge of static electricity. A discharge from a pd as low as 50 volts can cause component degradation and, since the effect is cumulative, repeated small discharges can lead to ultimate failure. It is therefore imporant that sensitive devices, such as LNBs, should be serviced in a workshop where static electricity can be controlled much better than on site. Bi-polar transistors are rather more robust in this respect.

The basic work station should provide for operators, work bench, floor mat, test equipment and device under service to be at the same electrical potential. Operators should be connected to the work bench via a wrist strap. Since static electricity is often generated by friction between dissimilar materials, the operator's clothing needs to be considered. The wearing of wool and man-made fibres such as nylon creates considerable static. One very useful garment is a smock made of polyester fabric, interwoven with conductive carbon fibres. This has through-the-cuff earthing. The use of a compressed air blast to clean down boards can actually generate static. This can be avoided by the use of an ionised air blast.

Static-sensitive components should be stored in conductive film or trays until required and then handled with short-circuited leads until finally in circuit. Soldering iron bit potentials can also be troublesome unless adequately earthed.

The work bench surface should be clean, hard, durable and capable of dissipating any static charge quickly. These static-free properties should not change with handling, cleaning/rubbing or with ambient humidity.

The use of an ionised air ventilation scheme can be an advantage. Large quantities of negatively and positively charged air molecules will quickly neutralise unwanted static charges.

11.7   SERVICING THE LNB

This section of the receiver system can be the most difficult to repair. Fortunately, to date, these units are proving to be very reliable. Servicing of S/L Band units is probably no more complex than repairing the tuner units of UHF television receivers. However, as the operating frequencies rise, both the cost of the test equipment and the level of expertise needed increase in proportion. Manufacturers' service departments can usually justify the necessary spectrum analyser and signal generator needed for signal tracing and circuit alignment. However, the smaller department is left with the choice of returning faulty units to the manufacturer, or using its inherent skill and ingenuity to provide a local service.

Each LNB is hermetically sealed and weather-proofed and the wave guide aperture is closed with a window that is transparent to signals. This may be made of glass or a plastic such as Mylar or Terylene. The latter materials can degrade and crack with aging, probably due to exposure to ultra-violet radiation, and therefore need periodic replacement. It is important to keep moisture out of any wave-guide-type structure, as corrosion gives rise to surface roughness, which increases attenuation.

The covers of most units are screwed in place and include gaskets that are not only part of the weather-proofing but are also RF conductive. These must be treated with care or replaced if a unit is opened for service work. Some units are enclosed in a Duralumin tube, one end of which is spun sealed. If it is thought economically viable, these can be opened by carefully removing the spun edge with a small chisel. The unit can be resealed, after servicing, using a silver-loaded epoxy adhesive. This material is very suitable for metals, and has a resistivity of about 500 $\mu\Omega$.cm, which ensures RF sealing. The curing time of these adhesives can be improved by raising the temperature to about 50°C.

It should be noted that when the covers of the LNB are removed, the wave-guide-like structure is changed. This gives rise to a change of circuit conditions so that the performance may change as a result. Also, it is not easy to inject or extract signals from a microwave circuit board without disturbing the operating conditions. Such coupling problems may be alleviated by using wave-guide-to-coaxial-cable adaptors. These consist of a short section of guide, blanked off at one end and containing a pick-up probe.

Signal generators such as the Marconi 6150 series, which are capable of providing both AM and FM signals, cover the frequency range of 1 to 18 GHz and are suitable for workshop use. The Marconi 6061A Gunn Device Oscillator, covering 8 to 11.5 GHz, is readily portable and suitable for on-site use. Gunn devices are very suitable for portable operation as they are capable of producing up to 10 mW of output power from battery supplies. However, if used as a resonant cavity oscillator, Gunn devices tend to have a power output that is frequency dependent, due to variation of circuit Q factor.

Because of the restricted bandwidth covered by each LNB (much less than 1 octave), the distortion component of a lower-frequency signal generator can in an emergency be made use of. A signal source may be stated to have a total harmonic content of typically −40 dBc (40 dB below the carrier or fundamental level). This distortion will predominantly be third order, and might have a level of −50 dBc. With the carrier set to 10 mW, the third harmonic level would be 0.1 μW, and this is of the same order as the RF input signal to a Ku Band LNB. Thus a C Band signal generator might, with care, be used to provide a Ku Band input.

The LNBs used for data services are sometimes separated into low-noise amplifier plus low-noise converter stages, and housed in separate casings. To some extent, this eases the servicing problems. However, the data services operate SPSC, relatively narrow band. This imposes additional restrictions on such parameters as oscillator drift and phase noise.

Each LNB is supplied by power over the coaxial feeder cable of typically 15 to 20 volts. There will be, contained within the unit, two voltage stabiliser circuits to provide collector/drain voltages and gate bias. This part of the circuit should include a transient suppressor for protection, typically a PN silicon device capable of responding to an overvoltage in about 1 picasecond. The first RF stage will generally be biased for a low-current/low-noise state, whilst succeeding stages will be biased for high gain.

The transistors used in the LNBs are very expensive, and any that are found to be faulty should be replaced by identical types. The device parameters actually form part of the circuit tuning. Resolder components with care: too much solder on a joint can affect its RF performance. Internal RF interstage screens should not be disturbed unless absolutely necessary. If they need to be removed, then silver-loaded epoxy adhesive can be very effectively used in the refitment.

Soldering to ceramic-based circuit boards can be difficult. The heat loss from the soldering iron bit is greater than that experienced with lower-frequency boards. To avoid static electricity problems, the soldering iron bit should be earthed direct to a convenient earthy connection on the LNB.

The system diagram of the test and alignment bench developed for the repair of Ku band LNBs is shown in Fig. 11.8 (courtesy of MCES Ltd (9)). The test signal source is provided from the sweep generator via a microwave relay to a pair of waveguides which are provided to meet the needs of two different types of waveguide coupling. The waveguides are also provided with rotational flanges to cater for both horizontal and vertical polarisations. The sweep generator may operate in its basic mode or as a spot frequency generator for alignment purposes. The generator output level is capable of being varied over the range −75 dBm to 7 dBm, but the typical amplitude used for basic alignment is in the order of −60 dBm. This ensures that LNBs are tested for sensitivity with a relatively low level input. Increasing this level allows such parameters as third-order intercept and intermodulation products to be evaluated.

The detector stage routes the IF output from the LNB under test back to the display section. This consists of a scalar network analyser to monitor the overall gain across the passband and a spectrum analyser to test for harmonic distortion, instability and spurious responses. The frequency counter is included so that the local oscillator frequency can be accurately set to within 100 KHz in 10 GHz.

The noise figure meter is capable of working with LNBs producing outputs up to 1800 MHz and measuring and displaying gain and noise figures to within 0.01 dB. This instrument provides a standard reference

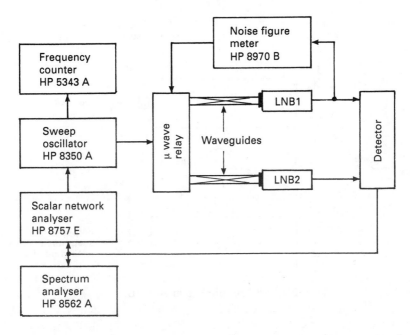

Fig. 11.8 LNB test bench (courtesy of MCES Ltd and Hewlett Packard Ltd).

noise level input to the LNB under test via the relay and waveguide. The IF output is routed back as shown and the noise contribution made by the LNB (the excess noise ratio) obtained by subtraction of the original noise level from the new measured value. Because the temperature of the waveguide can affect the noise figure this is continually monitored to allow correcting data to be entered into the final calculation.

A typically good Ku band LNB after repair would be expected to yield a gain of at least 55 dB, flat over the passband to within 3 dB, and with a noise figure of less than 1.5 dB.

Just as the LNB can be used as a front end to a spectrum analyser or alignment meter, a working LNB can in an emergency be pressed into service as an aid to fault finding. The basic arrangement is shown in Fig. 11.9. The inputs to the two LNBs are provided via wave guide adaptors and both are powered through their IF outputs. However, the test LNB can be powered from a spectrum analyser (AVCOM PSA-35) or an antenna alignment meter. The signal input to the test LNB is via a coaxial cable terminated in a small loop aerial of about 1 cm diameter, or a dipole probe of about 1 cm length. The presence of signal in the faulty LNMB can be detected by probing along the circuit board with the aerial, the point at which the signal is lost indicating the faulty stage. Fault finding in the IF section of the LNB then follows a procedure typical of UHF TV servicing.

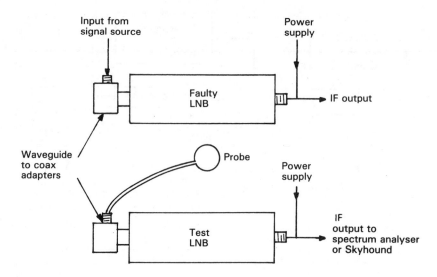

Fig 11.9 Fault-finding in an LNB.

11.8 SERVICING THE MAIN RECEIVERS

An Overview

The pre-demodulator stages of receivers for either analogue or digital services will incorporate a significant amount of analogue signal processing. This section may be either single or double conversion superhet based, depending on the system operating frequency range. The well tried and tested method of signal injection/signal tracing can be used to isolate a faulty stage. For Ku Band systems, the first IF will be very high and the circuit often composed of discrete components. Any repair or component replacements in this stage are thus likely to involve re-alignment using a swept frequency signal generator and spectrum analyser. This operation may be particularly critical with certain TVRO receivers, where the IF bandwidth is preselectable for different services. The work of fault finding in analogue stages is generally well documented (4). Post-demodulator processing may be either analogue or digital, the latter being well described in the literature (5).

Faults causing gain reduction in pre-demodulator stages, lead to a poor S/N ratio in analogue systems, the cause of which can readily be located by signal tracing. However, in digital services the same effect leads to an increase in bit errors, and the cause of this is not so easily traced.

The digital equivalent of an S/N ratio measurement can be the *eye height* measurement. If a data signal is applied to the vertical input of an oscilloscope whose timebase is locked in a particular way to the data rate, the resulting display becomes a series of superimposed signal transitions (6). Figure 11.10 represents such a display and shows how an 'eye pattern' is formed. The height H and width W of the eye are a function of the slopes of the data transitions; these, in turn, are dependent upon the degree of noise and distortion to which the signal has been exposed.

The eye height is defined as a ratio: the difference in levels between worst case 0s and 1s, to the difference in levels between a long sequence of 0s and 1s. From Fig. 11.10, this can be expressed as:

$$H = h/b \times 100\% \qquad (11.2)$$

The eye width W and jitter J are also significant in the interests of a low error rate. The greater the width of the eye, the lower will be the error rate after the resampling process. W is defined as the ratio of eye width to data bit period. From Fig. 11.10:

$$W = d/T \qquad (11.3)$$

Jitter, which is chiefly due to inter-symbol interference, also affects the error rate and can be evaluated as:

$$J = T(1 - W) \qquad (11.4)$$

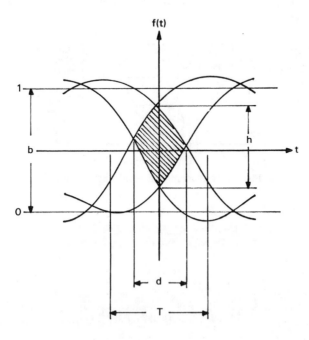

Fig. 11.10 Details of 'eye' display.

The principle of an eye pattern display unit is described with the aid of Fig. 11.11(6). A clock signal is derived from the data signal and its frequency divided by four. This signal is then low-pass filtered to obtain a sinusoid at 0.25 of the bit rate. This sinusoid and the original data stream are applied to the X and Y inputs, respectively, of an oscilloscope. By adjusting the phasing of the sinusoid, the displayed Lissajous figure becomes an 'eye'.

As an alternative, an oscilloscope can be used directly to provide an assessment of the eye height. If the raw data is applied to the Y input and the timebase set to run at a much lower frequency, the data transitions will be displayed in the manner shown in Fig. 11.12. Overshoots and undershoots generated by signal distortion give rise to the two brighter bands, which give the method its descriptive title – the 'tram-line' method. A reasonable approximation to the eye height can be obtained as:

$$a/b \times 100\% \tag{11.5}$$

When servicing a digital receiver, a before and after check of the eye height on the just demodulated signal can give a useful assessment of any improvements made. In general, a good eye height value would be around 75%. In practice, most digital receivers will work, with very few errors, down to an eye height of about 25% before the system crashes.

Test equipment required for servicing in the digital control section includes logic probes, frequency meters/counters and oscilloscopes. It is

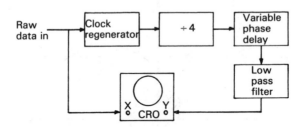

Fig. 11.11 Principle of 'eye' display unit.

Fig. 11.12 'Tram-lines' method of data display.

important to point out here that, even when displaying a relatively low bit rate signal, the CRO bandwidth should be sufficient to encompass at least the tenth harmonic, if the Y amplifiers are not going to further distort the data signal.

Analogue Section

The front end of these receivers may involve one or more stages of frequency conversion, prior to demodulation. Faulty stages can be readily isolated using the well-tried method of signal tracing/injection. Loss of gain in these IF stages leads to a worsening of S/N ratio for analogue services and an increase in bit error rate on digital services. For systems like the TVRO receivers, the first IF will typically be tunable over the range 950 to 1750 MHz, with the circuits consisting of discrete components. Faulty component replacement in such stages will involve checks of alignment, a swept frequency signal generator and spectrum analyser being almost obligatory to obtain satisfactory results. Second or subsequent IF stages are commonly based around integrated circuits, which introduce their own particular servicing problems. The first line of approach should be the comparison between input and output signals, followed by a check of dc levels. Take particular note of the pins on which identical voltages exist: these can point to short circuits which may be either on the chip or in an external component. If the chip is socket mounted, its removal will usually remove the element of doubt. If not, and if disconnecting the external feeds does not clear the way, then the chip may need to be treated as described in Section 11.5. After any servicing in the IF stages, it is important to check that gain and bandwidth are maintained across the tuning range.

Digital Section

The equipment needed for fault tracing in this section depends on the nature of the data signal. Whilst the signal is in serial form, it can be traced with an oscilloscope and its quality evaluated via an eye height measurement. When the data is in a parallel form, a logic analyser is more appropriate. For the system control stages, a logic probe will be found to be most effective. Problems related to frequency synthesis tuning systems should lead to an investigation first of dc levels, then of an accurate measurement of the main clock signal and its divided versions. Miscontrol action can arise because corrupted data has been stored in the system control EPROM. A useful starting point in such cases is to switch off, and then reprogram the tuning system.

Fault-finding in the Microprocessor Control Section

When a microprocessor control system fails, the parallel nature of its data organisation and the high degree of interdependence between the inputs and outputs, can make the task of fault finding difficult. All the elements of the system are interconnected via the data, address and control busses, so that a fault in one section can be reflected into several other areas. Typical bus line faults include short circuits between adjacent lines, or one or more lines jammed at logic 0 or 1 due to a short circuit to earth or positive supply rail.

Various test methods are described in the literature (7), some of which are only suitable for a large service organisation.

The logic analyser is the digital equivalent of the oscilloscope, but with parallel inputs. Modern instruments are available that will store as many as a hundred 32 bit words of data. These may be displayed, either as binary or hexadecimal characters, or as a series of waveforms, representing the high/low activity states of the various lines.

When a prototype system has been constructed and proved to function correctly, it can be made to execute a sequence of instructions in a repetitive manner. By monitoring each node of the system, data can be acquired about the correct logical activity. *Signature analysis*, as developed by Hewlett Packard Ltd, converts each node data into a unique four-character hexadecimal code that represents the 'signature' at a correctly functioning node. System documentation is then prepared, which includes a circuit diagram with all the correct node signatures appended. In servicing, a faulty component can then be identified as a device that produces an error output from correct input signatures.

Portable diagnostic aids based on a PROM that carries a stored test program may be plugged into a system test socket, or clipped 'piggy-back' fashion to the microprocessor; they can be valuable field service tools. Such devices are operated in some prescribed manner, so that the various elements of the system may be tested. Their status may be displayed either on LED/LCD, seven-segment indicators or, as in the case of a TV system, directly on the receiver screen.

Even without such aids, fault finding is still possible (8). The microprocessor IC is normally mounted in a socket, so that it can be removed to enable the bus lines to be checked for short circuits. An adaptor socket that isolates the microprocessor from the data bus can be interposed between this IC and its own socket. Forcing the microprocessor into a non-operation (NOP) or *free-running* mode causes it to issue every possible address value, in a continuous cycle. By comparing the pulse repetition frequency on adjacent lines, using a double-beam oscilloscope, the 2:1 relationship between correctly functioning lines can quickly be established.

The data bus lines can easily be checked for short circuits by the use of a logic probe. To examine its activity, the microprocessor needs to be forced into a *single-step* mode. It then becomes possible to check the validity of data transfers.

The status of the system clock waveform, or any derivatives, can easily be checked using an oscilloscope or logic probe.

Once the control section of the system has been proven to function correctly, attention can be directed towards the other elements in a systematic manner.

Unattended Operation

Some systems, such as those providing cable TV head-end feeds, are designed for unmanned operation. On these, a montly maintenance check is advisable. This can help considerably to reduce the incidence of service failure. The most important checks should include:

- Video and Audio channel outputs are noise free.
- AFC Status – if this is drifting towards the edge of its hold range, a readjustment might avoid a service failure.
- Relative signal levels – variation in such levels may give an indication of component aging, receiver drift or slight antenna misalignment problems due to weather conditions.
- System running temperature and installation weather proofing.

REFERENCES

(1) Watts, W. (1986) *Code of Practice for the Installation of Satellite Television Receiving Antennas*. London: Confederation of Aerial Industries.
(2) Frodsham, S., Commercial Chemicals Division, 3M UK Ltd, Bracknell, UK. Private communication to author.
(3) Cato, R.I., Multicore Solders Ltd, Hemel Hempstead, UK. Private communication.
(4) Cole, H.A. (1983) *Basic Colour Television*. Aldershot, UK: Gower Publishing Co. Ltd.
(5) Fisher, R. (1984) *Servicing Digital Circuits in TV Receivers*. London: Newnes Technical Books Ltd.
(6) Bennet, W.R. and Davey, J.R. (1965) *Data Transmission*. London: McGraw Hill.
(7) Day, S. (1981) 'Techniques for Fault finding in Microprocessor-Based Systems'. *I.B.A. Technical Review No. 15*. Independent Broadcasting Authority, UK.
(8) Williams, G.B. (1986) 'Simple Test Equipment for Microcomputers'. *Electronics and Wireless World*. Business Press International Ltd.
(9) Ayriss, J.F. and Glenton, J.A. M/C Colour Engineering Services Ltd, Manchester, UK. Private communication to author.

Appendix 1

A1.1 GEOSTATIONARY ORBITS

The gravitational force of attraction between two bodies is proportional to the product of their masses M, and inversely proportional to the square of the distance r between their centres of mass, the constant of proportionality being G, the Universal Gravitational Constant of 6.67×10^{-11} N m^2/kg^2.

Thus the gravitational force acting between a geostationary satellite and the earth (see Fig. A1.1) is given by:

$$F_g = GM_eM_s/r^2 \text{ Newtons} \tag{A1.1}$$

where M_e and M_s are the masses of the earth and satellite respectively
($M_e = 5.98 \times 10^{24}$ kg).

The force acting on a body, constrained to circular motion is proportional to its mass, the radius of the circular path and the square of the angular velocity. Thus the force acting on a geostationary satellite against gravity is given by:

$$F_c = M_sr\omega^2 \text{ Newtons} \tag{A1.2}$$

where r is the radius of the circular path and ω is the angular velocity.

The satellite continues in circular orbit around the earth when their separation and velocity are such that equations A1.1 and A1.2 are equal, so that:

$$GM_eM_s/r^2 = M_sr\omega^2 \text{ and}$$
$$r^3 = GM_e/\omega^2$$

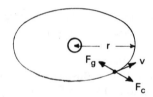

Fig. A1.1 The geostationary orbit.

Now, since the satellite is geostationary, its angular velocity ω is the same as that of the earth, 2π radians per sidereal day.

$$\omega = 2\pi/(23.9345 \times 60 \times 60) \text{ rad/sec}$$
$$= 7.30 \times 10^{-5} \text{ rad/sec (approx.)}$$

$$r^3 = (6.67 \times 10^{-11} \times 5.98 \times 10^{24})/(7.3^2 \times 10^{-10})$$
$$r = 42143 \text{ km (approx.)}$$

Assuming a mean earth radius of 6378 km, the satellite average height above the equator is 35765 km.

$$\text{Satellite velocity} = 2\pi(42143 \times 10^3)/(23.9345 \times 60 \times 60),$$
$$= 3.073 \text{ km/sec (approx.)}$$

A1.2 ELLIPTICAL ORBITS

Satellites using these orbits, shown in Fig. A1.2, obey Kepler's law of planetary motion fairly accurately. Rephrased for satellite applications these state:

(1) The satellite moves in an elliptical orbit with the centre of the earth at one focus.
(2) The radius vector sweeps out equal areas in equal time.
(3) The square of the period of revolution is proportional to the cube of the semi-major axis, giving rise to the equation:

$$T = 2\pi \sqrt{\frac{a^3}{GM_e}} \tag{A1.3}$$

where T is the orbit period in seconds, a is a semi-major axis in metres,

$$M_e = 5.98 \times 10^{24} \text{ kg.}$$
$$G = 6.67 \times 10^{-11} \text{ Nm}^2/\text{kg}^2$$

The velocity of the satellite is greatest at the perigee and least at the apogee. For this reason, the satellite is in communications range for fairly long periods at the greatest distance from earth. Using values taken from

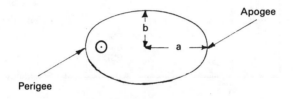

Fig. A1.2 The elliptical orbit.

the OSCAR-10 satellite (Orbiting Satellite Carrying Amateur Radio) gives an orbital time of approximately 11.66 hrs.

Semi-major axis	26100 km
Semi-minor axis	20800 km
Apogee	35500 km
Perigee	3960 km

Observations taken from satellites in these orbits show that equation A1.3 needs only a small correction, to account for the earth's elliptical shape and anisotropic density.

A1.3 FM/AM NOISE ADVANTAGES

Noise power proportional to v^2

Noise power (AM) $= \int_0^{f_1} 1^2.\, df = f_1$ watts.

Noise power (FM) $= \int_0^{f_1} v^2.\, df.\ \left[v = 1 \times \dfrac{f_1}{f^2},\ \therefore v^2 = \left(\dfrac{f_1}{f_2}\right)^2 \right]$

$$= \int_0^{f_1} \left(\frac{f_1}{f_2}\right)^2 .df \quad (f_2 \text{ is a constant})$$

$$= \frac{1}{f_2^2} \left[\frac{f_1^3}{3}\right] = \frac{f_1^3}{3f_2^2} \text{ watts.}$$

Ratio AM noise/FM noise

$$= \frac{f_1 \times 3f_2^2}{f_1^3} = 3\left(\frac{f_2}{f_1}\right)^2 = 3M^2 \left[M = \frac{f_2}{f_1} \text{ , the deviation ratio} \right]$$

Factor of improvement, S/N ratio $= 3M^2$

When $M = 5$, the noise improvement $= 10 \log. 25 = 18.75$ dB.

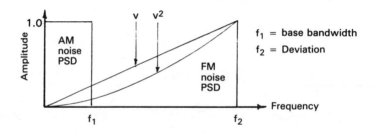

Fig. A1.3 AM/FM noise power spectral density.

A1.4 DE-EMPHASIS

$$v_0 = \frac{i}{j\omega c}$$

$$i = \frac{v_i}{R + \dfrac{1}{j\omega c}}$$

$$v_i = iR + \frac{i}{j\omega c}$$

$$\frac{v_o}{v_i} = \frac{\dfrac{1}{j\omega c}}{R + \dfrac{1}{j\omega c}} = \frac{1}{1 + j\omega CR} = \frac{1}{1 + j\omega t} \quad [t = CR]$$

$$\left| \frac{v_o}{v_i} \right| = \frac{1}{(1 + \omega^2 t^2)^{\frac{1}{2}}} \text{ or } \frac{1}{\sqrt{(1 + \omega^2 t^2)}}$$

$$20 \log \left| \frac{v_o}{v_i} \right| = 20 \log (1 + \omega^2 t^2)^{-\frac{1}{2}} = -10 \log (1 + \omega^2 t^2)$$

to give the circuit response at any frequency, for example, the response at 10 KHz with a time constant t = 50μS is:

$$-10 \log (1 + (2\pi)^2 \times 10^8 \times 50^2 \times 10^{-12}) \text{ dB}$$
$$= -10 \log (1 + 9.87) \text{ dB}$$
$$= -9.95 \text{ dB}.$$

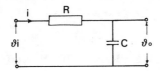

Fig. A1.4 De-emphasis network.

J.17 Pre-emphasis

This standard, that was originally designed for use with sound programme links, is now often applied to satellite delivered sound channels. As shown in Fig. A1.5, the response curve has maximum effect over the mid-range

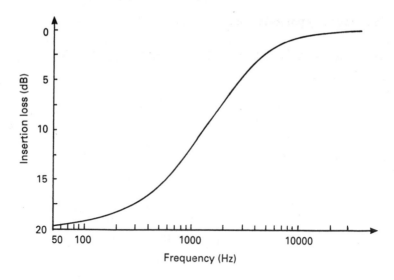

Fig. A1.5 J17 Pre-emphasis characteristic.

frequencies, and this is beneficial when handling modern music. The levelling off at HF is also helpful in minimising sibilant effects that are often present with single order pre-emphasis circuits.

The response curve represents an insertion loss given by;

$$10 \log \frac{75 + (\omega/3000)^2}{1 + (\omega/3000)^2} \; \text{dB}$$

where $\omega = 2\pi f$ and f is the corresponding frequency.

The curve is practically symmetrical about 1.2 KHz and provides about 6 dB boost relative to LF at 800 Hz.

The receiver de-emphasis circuit obviously needs to have a complementary response.

A1.5   SIGNALS IN THE PRESENCE OF NOISE

This section is not intended to be definitive, but is offered more in the revisionary sense. The reader should refer to such texts as are listed in the bibliography (1,2,3) for a more detailed analysis of signal behaviour in the presence of noise.

White noise has a normal or Gaussian distribution; that is, it has an amplitude distribution described statistically by the curve of Fig. A1.6. The curve is drawn for the function:

$$y = \frac{1}{\sqrt{(2\pi)}} e \left(-\frac{x^2}{2} \right)$$

The following points should be noted:

(1) When $x = 0$, $y = \dfrac{1}{\sqrt{2\pi}}$ or approximately 0.4.

(2) The curve is symmetrical about the y axis.

(3) The area under the curve is unity.

(4) y approaches zero as x tends to infinity.

(5) For this normal distribution, the area under the curve represents the probability that x ($P(x)$) falls within a range of values. For example:

$$P(x) \; -\infty < x < +\infty = 1,$$
$$P(x) \text{ is negative} = P(x) \text{ is positive} = 0.5,$$

In general, the probability that x lies between a and b is given by the area bounded by a, b and the curve. For convenience, the probability values are tabulated.

The following is an extract from tables such as those given in reference (1).

x	0	0.5	1.0	1.5	2.0	2.5

Area between

0 and x	0	0.1915	0.3413	0.4332	0.4772	0.4938

Thus:

Probability that x lies between 0 and 1 $= 0.3413$
Probability that x lies between 1 and 2 $= 0.1359$
Probability that x is greater than 1.5 $= 0.5 - 0.4332$
$= 0.0668$

Because of symmetry, the mean (μ) or average value of the function is zero. In a practical situation this is unlikely, and it is important to know the dispersion or deviation from the mean that the values take. The most effective parameters in these cases are the *variance* (σ^2) and the *standard deviation* (σ). The variance of the distribution is the average of the sum of the squares of the deviation from the mean value, whilst the standard

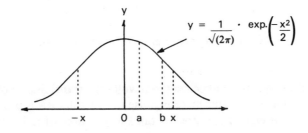

$$y = \frac{1}{\sqrt{(2\pi)}} \cdot \exp.\left(-\frac{x^2}{2}\right)$$

Fig. A1.6 The normal distribution.

deviation is simply the square root of the variance. For the normal distribution with $\mu = 0$, $\sigma^2 = 1$ and $\sigma = 1$.

In the case where the mean is non-zero, the probability is given by the area under the curve of the function:

$$y = \frac{1}{\sqrt{(2\pi\sigma)}} \cdot e^{\left(\frac{-x^2}{2\sigma^2}\right)}$$

which is much less easy to use. For such cases, other functions are tabulated (3).

The probability that x lies between $-x$ and $+x$ is given by the *error function* (erf) where:

$$\text{erf}(x) = 2\left[\frac{1}{\sqrt{\pi}} \int_0^x \exp(-y^2) \, dy\right]$$

The probability that x lies outside this range is given by the *complementary error function* (erfc) where:

$$\text{erfc}(x) = 1 - \text{erf}(x)$$

It is shown (2, 3) that the probability of error P_e is given by:

$$P_e = \tfrac{1}{2}\text{erfc}\left[\frac{x}{\sigma\sqrt{2}}\right]$$

Example (Fig. A.1.7)

A binary baseband transmission system operates at 50 Kb/s. Logic 1 is

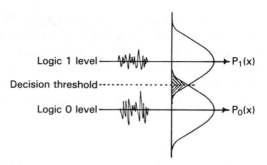

Fig. A1.7 Diagram of error region for example.

represented by $+1$ volt, logic 0 by -1 volt, the transmission of ones and zeros are equiprobable and there is noise of 0.2 volts RMS present.

Probability of error P_e = area of tail $P_1(x)$ + area of tail $P_0(x)$.

$P_e = 2 \times$ area one tail

$$= \tfrac{1}{2} \text{ erfc} \left[\frac{V}{\sigma \sqrt{2}} \right] = \tfrac{1}{2} \text{ erfc} \left(\frac{5}{\sqrt{2}} \right) = \tfrac{1}{2} \text{ erfc. } 3.536$$

$P_e = \tfrac{1}{2}[1 - (\text{erf } 3.536)]$

$\quad = \tfrac{1}{2}[1 - 0.999999425]$

$\quad = 2.875 \times 10^{-7}$ errors/sec/per bit

At 50kb/s $\text{BER} = 2.875 \times 10^{-7} \times 50 \times 10^3$

$$= 0.014375 \text{ errors/sec.}$$

or approximately 1 error per 70 seconds.

If the noise level is doubled to 0.4 volts RMS, the probability of errors increases.

$P_e = \tfrac{1}{2} \text{ erfc} \left(\frac{2.5}{\sqrt{2}} \right) = \tfrac{1}{2} \text{ erfc } 1.770$

$P_e = \tfrac{1}{2}[1 - 0.987691]$

$\quad = 6.155 \times 10^{-3}$ errors/sec/per bit

At 50 kb/s, $\text{BER} = 6.155 \times 10^{-3} \times 50 \times 10^3$

$$= 308 \text{ errors per sec.}$$

This shows the disastrous effect that noise can have, even on a digital signalling system.

REFERENCES

(1) Mosteller, F. *et al.* (1970) *Probability with Statistical Applications.* 2nd edn. World Student Series. London: Addison-Wesley Publishing Company.
(2) Schwartz, M. (1970) *Information Transmission, Modulation and Noise.* 2nd edn. International Student Series. London: McGraw-Hill Kogakusha Ltd.
(3) Betts, J.A. (1978) *Signal Processing, Modulation and Noise.* London: Hodder and Stoughton Ltd.

Appendix 2

A2.1 NOISE PERFORMANCE OF CASCADED STAGES

For the purposes of noise analysis, the noise generated within a system can be replaced by an equivalent auxiliary input of noise so that the system can be treated as ideal (non-noise-producing).

In Fig. A2.1, system and source are correctly matched so that the noise input is kTB watts. The system has a noise factor F and power gain G, so that the total noise output power is $N_o = FGkTB$ watts. If the system had been ideal, this would have been $N_o = GkTB$ watts. Therefore the noise power generated within the system is:

$$FGkTB - GkTB = GkTB(F - 1)$$

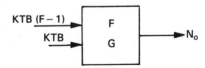

Fig. A2.1 Noise performance of single stage.

This is equivalent to providing the system with an auxiliary noise input of kTB(F − 1) watts.

This concept can be extended for two or more systems in cascade as depicted in Fig. A2.2, where the total noise output is:

$$
\begin{aligned}
N_o &= kTBG_1G_2 + kTB(F_1 - 1)G_1G_2 + kTB(F_2 - 1)G_2 \\
&= kTBG_1G_2 + kTBF_1G_1G_2 - kTBG_1G_2 + kTB(F_2 - 1)G_2 \\
&= kTBG_1G_2F_1 + kTB(F_2 - 1)G_2
\end{aligned}
$$

Noise output due to input alone $= kTBG_1G_2$

$$F_o = \frac{\text{Total noise power at output}}{\text{Noise power at output due to input alone}}$$

$$= \frac{kTBG_1G_2F_1 + kTB(F_2 - 1)G_2}{kTBG_1G_2}$$

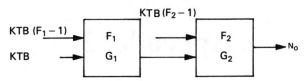

Fig. A2.2 Noise performance of cascaded stages.

$$= F_1 + (F_2 - 1)/G_1.$$

Similarly for three stages in cascade:

$$F_o = F_1 + (F_2 - 1)/G_1 + (F_3 - 1)/G_1 G_2 \qquad (A2.1)$$

Appendix 3

A3.1   THE UNIVERSAL ANTENNA CONSTANT

Consider two antennas R metres apart and directed towards each other as shown in Fig. A3.1, the gains and areas of each being G_1, A_1 and G_2A_2 respectively. A power of P watts is transmitted from antenna 1 towards antenna 2 so that the power flux density (PFD) at 2 is $PG_1/4\pi r^2$ watts/m². The total power received at antenna 2 is therefore $PG_1A_2/4\pi r^2$ watts.

Now exchange transmitter and receiver without changing any other parameter. The total power received at 1 is $PG_2A_1/4\pi r^2$ watts. Applying the reciprocity theorem:

$$PG_1A_2/4\pi r^2 = PG_2A_1/4\pi^2$$
$$G_1A_2 = G_2A_1 \text{ and}$$
$$G_1/A_1 = G_2/A_2, \text{ which is a constant.}$$

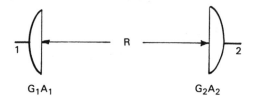

Fig. A3.1 Diagram for derivation of Universal Antenna Constant.

If this constant can be determined for just one antenna, then it can be applied to all antennas. It is shown in reference (1) that a small Hertzian dipole of length l, which is much less than half a wavelength, has a gain $G = 1.5$ and radiation resistance $R = 80(\pi l/\lambda)^2$ ohms. If such an antenna is placed in an electric field of E volts/m, it will collect an emf of $E.e_{jwt}.l$ volts. The maximum power that this will generate under matched conditions will be $(1/2(emf)^2)/4R$ watts.

$$\text{Maximum power} = (1/2) (El)^2/(4(80(\pi l/\lambda)^2))$$
$$= E^2\lambda^2/640\pi^2 \text{ watts.}$$

The power flow in the wave is $(1/2)E^2/Z_o$, where Z_o is the impedance of free space $= \sqrt{(\mu_o/\epsilon_o)} = 120\pi$ ohms. If the effective area of the receiving antenna is A_e square metres, the received power is:

$E^2 A_e/240\pi$ watts.

Therefore, $E^2 A_e/240\pi = E^2 \lambda^2/640\pi^2$ or

$A_e = 3\lambda^2/8\pi$.

For the Hertzian dipole:

$$G/A_e = 3/2 \times 8\pi/3\lambda^2$$
$$= 4\pi/\lambda^2, \text{ the universal constant.}$$

The gain G is thus $4\pi A_e/\lambda^2$.

A3.2 APPROXIMATE BEAMWIDTH FOR PARABOLIC REFLECTOR ANTENNAS

Referring to Fig. A3.2, where each small elemental area dA is matched to a corresponding area dA' spaced by a radius of half the diameter D, assume uniform illumination of the dish at wavelength λ and uniform phase. For the half power, direction $\theta/2$ has to be such that the contributions to the transmitted power by dA and dA' are quadrature, i.e. $x = x'$ and $x + x' = \lambda/4$.

Sin $\theta/2 = (\lambda/8)/(D/4) = \lambda/(2D)$. For high-gain antennas $\theta/2$ is small, so that: sin $\theta/2 = \theta/2 = \lambda/(2D)$ or
$\theta = \lambda/D$ radians

Or for θ in degrees, the half power beamwidth $= 57.3\lambda/D$.

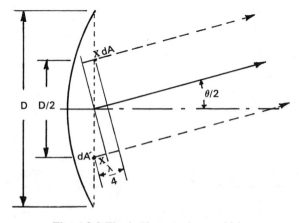

Fig. A3.2 The half-power beamwidth.

A3.3   AZIMUTH/ELEVATION ANGLES BY CALCULATOR

A calculator with trigonometrical functions and this algorithm can quickly produce the azimuth/elevation angles. For the northern hemisphere, the convention is to use positive angles for satellites and ground stations west of the Greenwich meridian. (Easterly locations use corresponding negative values.)

(1)   Find the angle WEST (the longitude difference). Satellite °W – station °W.
(2)   Find $X = (\cos \text{WEST}° \times \cos(\text{station latitude}°))$
(3)   AZIMUTH = arc tan (tan WEST/sin station latitude)
        Add 180° if satellite is west of Greenwich meridian.
(4)   Calculate $Y = \sqrt{1 + K^2 - 2KX}$, where $K = 6.608$, the distance between the satellite and the earth centre in terms of earth radii.
(5)   ELEVATION = arc cos $((1 - KX)/Y) - 90°$
(6)   Calculate range $Z = 6378 \times Y$ km.

Note: for latitudes greater than 81°, $X < 0.15$; this indicates that the satellite is below the horizon.

A3.4   CALCULATION OF ANGLE OF ELEVATION α FOR SIMPLE POLAR MOUNTS

With reference to Fig A3.3, and using the cosine rule:

$$\cos \beta = \frac{(r+h)^2 + r^2 - d^2}{2r(r+h)} \quad \text{so that}$$

$$d = \sqrt{(r+h)^2 + r^2 - 2r(r+h)\cos \beta}$$

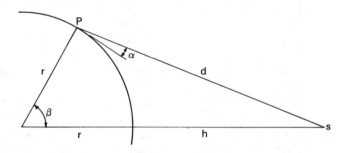

Fig. A3.3 The simple polar mount calculations.

Using the sine rule:

$$\frac{r+h}{\sin(90+\alpha)} = \frac{d}{\sin \beta} = \frac{\sqrt{(r+h)^2+r^2-2r(r+h)\cos \beta}}{\sin \beta}$$

$$\frac{(r+h)\sin \beta}{\sqrt{((r+h)^2+r^2-2r(r+h)\cos\beta)}} = \sin(90+\alpha) = \cos \alpha$$

$$\alpha = \text{arc cos}\frac{(r+h)\sin \beta}{\sqrt{((r+h)^2+r^2-2r(r+h)\cos\beta)}}$$

Substituting for the constants $r=6378 \times 10^3$ and $h=35765 \times 10^3$, this simplifies to:

$$\alpha = \text{arc cos}\frac{1.82\sin \beta}{\sqrt{(3.38-\cos \beta)}} \tag{A3.4}$$

This function has been calculated (Table A3.1) and plotted. For values up to latitude (β) of 55°, the relationship is linear to within 1%. The angle of elevation can thus be derived from:

$$\alpha = 90° - 1.15385\beta \tag{A3.5}$$

Table A3.1

Latitude°	0	10	20	30	40	50	60	70	80
Elevation°	90	78.22	66.52	54.48	43.65	32.57	22.56	11.12	0

A3.5 MODIFIED POLAR MOUNT

In chapter 3, reference is made to the need to apply a declination correction to the elevation angle of the simple polar mount. This arises chiefly because of the earth-bound observer's view of the Clark Orbit. From an equatorial viewpoint, the orbit appears as a straight line, running due east to west and passing overhead. If the orbit could be viewed from either of the polar regions, it would be seen to be circular. At all latitudes between these two extremes, the orbit appears elliptical. The simple polar mount provides a tracking arc that is circular, so that it will only coincide with this apparent ellipse tangentially in a southerly direction. At the best, it will track only two points if the elevation angle is slightly offset. This may be acceptable if the services of only two satellites are to be used.

Figure A3.4(a) shows a plan view (not to scale) of the earth and the Clarke Orbit, through the equatorial plane, as seen from above the north pole. Point S represents a satellite positioned due south of an earth station, located at Q (T in Fig. A3.4 (c)). The antenna can be swung through 90° in both an easterly and a westerly direction, to point to two further satellites S_1

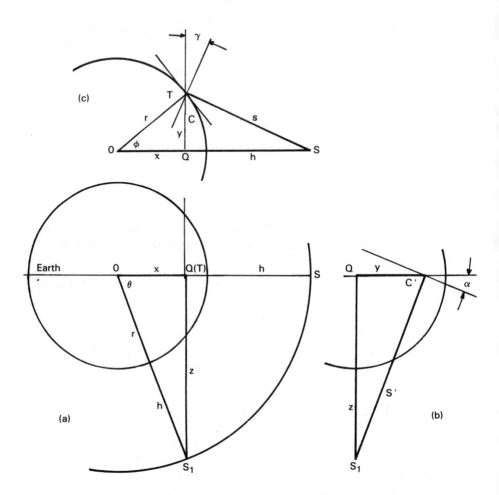

Fig. A3.4 Triangulation for offset polar mount analysis.

and S_2. (S_2 is in a mirror-image position relative to S_1, about the line OS.) Although these points are below the horizon, they are used in this simplified . analysis (2) to provide for three-point accurate tracking.

With reference to Fig. A3.4(b), the antenna is aimed at satellite S_1, with a declination angle of α. Swinging the antenna through an angle of 90° to point to satellite S causes the required angle α to increase by a small value. The new declination angle $\gamma = \alpha + \beta$, where β is the additional offset angle required to accurately track the three satellites. The angle β is never greater than 1°.

For the three triangles shown in Fig. A3.4, the angle ϕ is the earth station latitude and θ is the difference in longitude between S and S_1, and S and S_2.
From Fig. A3.4(c):

$$x = r \cos \phi$$
$$y = r \sin \phi$$
$$SQ = (r + h) - x$$

In triangle SQT:

$$C = \text{arc tan} \frac{(r + h) - x}{y}$$

$$= \text{arc tan} \frac{(r + h) - r \cos \phi}{r \sin \phi} \qquad (A3.6)$$

From Fig. A3.4(b):

In triangle $S_1 Q T_1$
$$C' = 90° - \alpha$$
$$= \text{arc tan} \frac{z}{y} \qquad (A3.7)$$

From Fig A3.4(a)

In triangle $S_1 Q O_1$
$$z^2 + x^2 = (r + h)^2$$
$$z = \sqrt{(r + h)^2 - r^2 \cos^2 \phi}$$

Therefore equation A3.7 becomes:

$$C' = \text{arc tan} \frac{\sqrt{(r + h)^2 - r^2 \cos^2 \phi}}{r \sin \phi} \qquad (A3.8)$$

and

$$\alpha = 90° - C'$$
$$= 90° - \text{arc tan} \frac{\sqrt{(r + h)^2 - r^2 \cos^2 \phi}}{r \sin \phi} \qquad (A3.9)$$

From Fig. A3.4(c)

$$\gamma = \alpha + \beta, \text{ so that}$$
$$\alpha + \beta + c = 90° \text{ or}$$
$$\beta = 90° - \alpha - c$$
$$= 90° - 90° + \text{arc tan} \frac{\sqrt{(r + h)^2 - r^2 \cos^2 \phi}}{r \sin \phi}$$

$$- \text{arc tan} \frac{(r + h) - r \cos \phi}{r \sin \phi} \qquad (A3.10)$$

Therefore

$$\beta = \arc\tan \frac{\sqrt{(r+h)^2 - r^2 \cos^2 \phi}}{r \sin \phi}$$

$$- \arc\tan \frac{(r+h) - r \cos \phi}{r \sin \phi} \tag{A3.11}$$

If the values for the constants r and h are inserted into equations A3.9 and A3.11, the graphs shown in Fig. A3.5 can be plotted. The due south elevation angle for the modified polar mount can then be obtained for any latitude from:

$$90° - (\alpha + \beta + \phi) \tag{A3.12}$$

This value provides tracking to an accuracy within 0.05°.

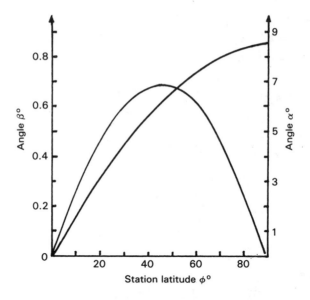

Fig. A3.5 Plot of equations A3.9 and A3.11.

REFERENCES

(1)   Connor, F.R. (1984) *Introductory Topics in Electronics and Tele-communications: Antennas.* London: Edward Arnold.
(2)   Birkill, S.J. Technical Director Satellite TV Antenna Systems Ltd, Staines, UK. Private communication to author.

Author's footnote: this analysis is due to an idea presented by Steve Birkill, for which I am particularly grateful.

Appendix 4

A4.1 QUANTISATION NOISE

Figure A4.1 represents the quantisation errors that are introduced when an analogue signal is converted into digital form. The peak-to-peak signal amplitude (P) is divided into M levels, each of step amplitude 'a'.

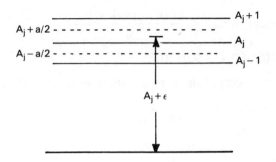

Fig. A4.1 Quantisation errors.

Statistically, any value to be quantised will fall in the band $A_j \pm a/2$, so that the mean square error $\overline{\epsilon^2}$ is given by:

$$\overline{\epsilon^2} = \frac{1}{a} \int_{-a/2}^{+a/2} \epsilon^2 \, d\epsilon$$

$$= \frac{1}{a} \left[\frac{\epsilon^3}{3} \right]_{-a/2}^{+a/2} = \frac{1}{a} \left\{ \left[\frac{(a/2)^3}{3} \right] - \left[\frac{(-a/2)^3}{3} \right] \right\}$$

$$= \frac{1}{3a} \left[\frac{a^3}{4} \right] = a^2/12$$

RMS error $= a/\sqrt{12}$.

Now the signal-to-quantisation noise power ratio $SQNR = p^2/(a^2/12)$; and since $P = aM$, $SQNR = 12a^2M^2/a^2 = 12M^2$, or:

$$SQNR_{dB} = 10.8 + 20 \log M. \tag{A4.1}$$

The larger the value of M, the greater the number of bits required to code each sample, and hence the wider the base bandwidth. The S/N ratio for binary signalling can be related to the bandwidth as follows:

$$M = 2^n$$

where n is the number of bits per sample, so that the

$$
\begin{aligned}
\text{SQNR} &= (10.8 + 20 \log 2^n)\text{dB} \\
&= (10.8 + 20n \log 2)\ \text{dB} \\
&= (10.8 + 6n)\ \text{dB}.
\end{aligned}
\tag{A4.2}
$$

The equation A4.2 representing the peak to peak signal to RMS quantisation noise adequately expresses the annoyance effect of quantisation noise on a signal such as video (varying dc level). For ac signals such as audio, the RMS signal to RMS quantisation noise is more appropriate.

The two ratios can be reconciled as follows:

$$\text{RMS signal voltage} = \frac{\text{peak to peak value}}{2\sqrt{2}}$$

$$20 \log (1/2\sqrt{2}) = -9.03\ \text{dB}$$

so that the ac version of signal to quantisation noise is often expressed as:

$$(1.76 + 6.02n)\ \text{dB}.
\tag{A4.3}$$

Appendix 5

A5.1 MICROWAVE FREQUENCY BAND CLASSIFICATION

As stated in chapter 1, the frequency ranges used for satellite communications are commonly known by the American Radar Engineering Standard. In Europe, however, another classification may be used, and this can give rise to some confusion.

Table A5.1 Frequency Ranges (GHz)

Band	American		European	
P	0.2	– 1.0	0.2	– 0.375
L	1	– 2	0.375	– 1.5
S	2	– 4	1.5	– 3.75
C	4	– 8	3.75	– 6
X	8	– 12.5	6	– 11.5
J	–	–	11.5	– 18
Ku	12.5	– 18	–	–
K	18	– 26.5	18	– 30
Ka	26.5	– 40	–	–
Q	–	–	30	– 47

Index